A BOOK OF
CURVES

A BOOK OF
CURVES

BY

E. H. LOCKWOOD

Senior Mathematics Master, Felsted School
Formerly Scholar of St John's College
Cambridge

CAMBRIDGE
AT THE UNIVERSITY PRESS
1963

CAMBRIDGE UNIVERSITY PRESS
Cambridge, New York, Melbourne, Madrid, Cape Town, Singapore, São Paulo

Cambridge University Press
The Edinburgh Building, Cambridge CB2 8RU, UK

Published in the United States of America by Cambridge University Press, New York

www.cambridge.org
Information on this title: www.cambridge.org/9780521055857

© Cambridge University Press 1961

First published 1961
Reprinted 1963
This digitally printed version 2007

A catalogue record for this publication is available from the British Library

ISBN 978-0-521-05585-7 hardback
ISBN 978-0-521-04444-8 paperback

CONTENTS

*

His Word accomplish'd the design

CHRISTOPHER SMART

*

PREFACE

Plane curves offer a rich and to some extent unexplored field of study which may be approached from a quite elementary level. Anyone who can draw a circle with a given centre and a given radius can draw a cardioid or a limaçon. Anyone who can use a set square can draw a parabola or a strophoid. Anyone who knows a few of the simpler propositions of Euclid can deduce a number of properties of these beautiful and fascinating curves.

In school they may be used to instruct and entertain classes at all levels. In a class of mixed ability some will pursue the theory while others continue with the drawing.

Teachers may use the book in a variety of ways, but it has been written also for the individual reader. It is hoped that it will find a place in school libraries, and will be used too by sixth-form pupils, whether on the arts or the science side, who have time for some leisurely work off the line of their main studies, time perhaps to recapture some of the delight in mathematics for its own sake that nowadays so rarely survives the pressure of examination syllabuses and the demands of science and industry.

The approach is by pure geometry, starting in each case with methods of drawing the curve. In this way an appreciation of the shape of the curve is acquired and a foundation laid for a simple geometrical treatment. There may be some readers who will go no further, and even these will have done more than pass their time pleasantly; but others will find it interesting to pursue the geometrical development at least to the point at which one or other of the equations of the curve is established. Those who have a knowledge of the calculus and coordinate geometry may prefer to leave the text at this point and find their own way, using as a guide the summary of results which will be found at the end of each chapter of Part I and some chapters of Part II.

In Part II the reader is encouraged to explore further for himself, using whatever resources are available to him. While some individual curves are briefly discussed, this part of the book is mainly concerned with methods by which new curves can be found. Those whose delight is in the drawing will find much to occupy them here; but deduction can often contribute both to the shaping of a curve and to a discovery of its properties.

My particular thanks are due to Mr A. Prag, mathematics master and Librarian at Westminster School, who has written the Historical Introduction and most of the shorter historical notes, a scholarly contribution without which the book

[vii]

would have been sadly incomplete. I am grateful also to my colleague at Felsted Mr P. Gant, who has read the manuscript in detail and has offered many helpful criticisms and suggestions; to Mr Alan Breese, who made the drawings for the full-page diagrams; and to Dr A. M. Winchester, whose photograph of the Pearly Nautilus fossil appears as the frontispiece. Finally, I express my gratitude to the Syndics and Staff of the Cambridge University Press: how much the book owes to them in its planning and production the reader can now judge.

FELSTED
August 1960

E. H. L.

HISTORICAL INTRODUCTION

Men were fascinated by curves and curved shapes long before they regarded them as mathematical objects. For evidence one has only to look at the ornaments in the form of waves and spirals on prehistoric pottery, or the magnificent systems of folds in the drapery of Greek or Gothic statues. It was the Greek geometers who began to study geometrically defined curves as, for instance, the contour of the intersection of a plane with a cone, or the locus of points reached one by one through a geometrical construction. The straight line and the circle could be drawn with very primitive instruments in one continuous movement, and so they were distinguished as 'plane' loci from the 'solid' conic sections. All other curves were *loci lineares*, i.e. just 'lines', 'curves'. Some curves were generated by the movement of mechanical linkages, or at least were imagined to be so generated: the spirals of Archimedes were of that type. A classification into 'geometrical' and 'mechanical' curves (which does not quite correspond to the modern use of those terms) became fixed when analytical geometry, in the seventeenth century, made it possible to distinguish with precision what we should now (following Leibniz) call *algebraic* and *transcendental* curves.

In his search for the true shape of a planetary orbit Kepler tried a variety of curves before he found that the ellipse gave the best fit. In the old Ptolemaic system the planets were supposed to describe paths which could be constructed by means of *epicycles* (i.e. by circles carried on other circles or spheres). Kepler altogether enjoyed playing with curves and invented a great number of names (usually those of some sort of fruit) for the solids of revolution generated by curves rotating about various axes.

When Cavalieri tried to explain his method of integration (*Cavalieri's Principle*) he was careful to use a really general type of curve, but he lacked the analytical method of description; later in the seventeenth century, Gregory and Barrow gave the rules of the calculus (as we should call it) in geometrical form by referring to simple monotone arcs. Thus already the individual curves were beginning to be lost in more general theory.

A powerful device was the creation of a new curve by the transformation of another, as for instance, a curve formed by drawing ordinates equal to the lengths of the subtangents of a given one. A simpler example was the drawing of a *conchoid* $r = f(\theta) + c$ for a given curve $r = f(\theta)$: then, if the tangent to the given curve were known, the tangent to the conchoid could immediately be constructed. Problems in optics led to *caustics*, i.e. envelopes of pencils of rays. But the greatest influence in the study of curves was, of course, the invention of the calculus, which not only secured the solution of problems on gradients, areas, and lengths of arcs, but unified the whole field of research. A great variety of mechanical problems could then be precisely formulated, as, for instance, to find the curve of 'quickest descent', the *brachistochrone*.

But the interest had shifted from the geometrical origin of the concept 'curve' to the analytical aspect: it was as a diagram of a 'function' that the curve appeared in the text-book, and the individuality of many famous members of the family was lost.

NOTATION

The following notation will be used, more particularly in the summary of results at the end of each chapter:

(r, θ) are polar coordinates.

t is a parameter for the parametric equations of a curve.

ϕ = angle between radius vector and tangent.

ψ = angle between initial line (or x-axis) and tangent.

s = arc-length, measured usually from $\theta = 0$ or $t = 0$.

p = perpendicular distance from origin to tangent.

ρ = radius of curvature.

A = area enclosed by a curve.

L = total length of a closed curve.

The letters P, P' will be used to name points on the curve which is being drawn; Q, Q' for points on a subsidiary line or curve; and q for a point which will eventually move towards and coincide with Q.

For the drawing of the curves, suitable dimensions will be suggested for each of the following sizes of paper:

Size 1	9 in. by 7 in.	23 cm. by 18 cm.
Size 2	13 in. by 8 in.	33 cm. by 20 cm.
Size 3	15 in. by 11 in.	38 cm. by 28 cm.

When it is necessary to specify which way up the paper is to be used, the suffixes P, for the 'portrait' (i.e. upright) position, or L, for the 'land-scape' position, will be added. Thus 'Paper 1_P' means 'Size 1, in the portrait position'.

Lines drawn across the paper from left to right will sometimes be referred to as 'horizontal' and lines drawn up the paper as 'vertical'.

* An asterisk will be used to mark the more difficult sections and exercises.

** A double asterisk will indicate work which, though not necessarily difficult, demands knowledge beyond the syllabuses of O-level 'Additional Mathematics'.

PART I

SPECIAL CURVES

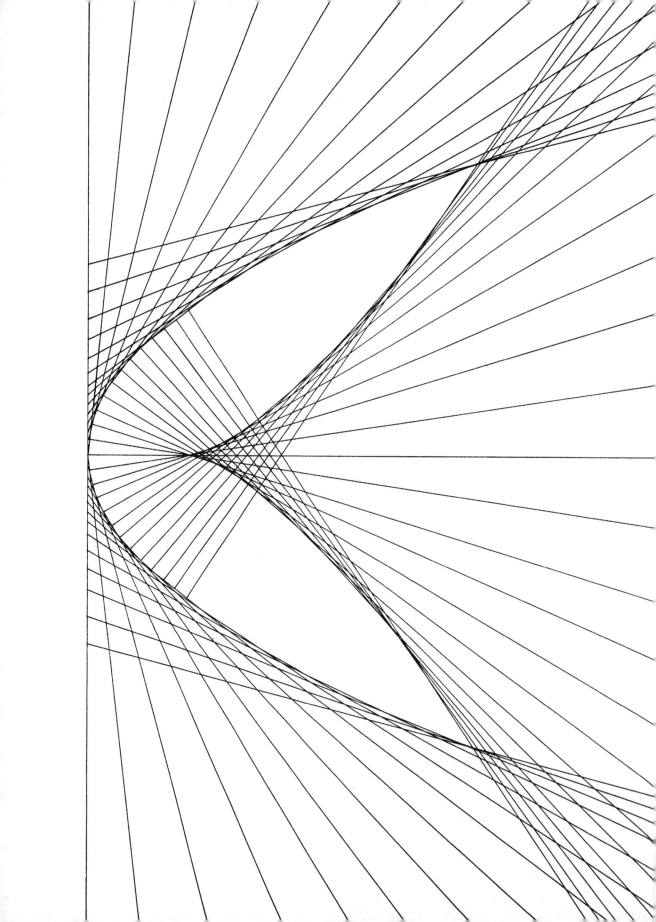

1

THE PARABOLA

To Draw a Parabola

Draw a fixed line AY and mark a fixed point S. Place a set square UQV (right-angled at Q) with the vertex Q on AY and the side QU passing through S (Fig. 2). Draw the line QV. When this has been done in a large number of positions, the parabola can be drawn freehand, touching each of the lines so drawn. The curve is said to be the *envelope* of the variable line QV (Fig. 1).

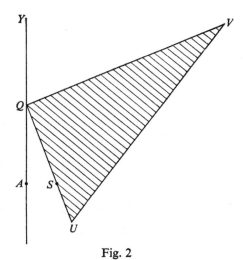

Fig. 2

Suitable Dimensions

With AY near to and parallel to the left-hand edge of the paper, the distance of S from AY should be approximately as follows:

$$
\begin{array}{lll}
\text{Paper:} & 1_P & \text{0·8 in. or 2 cm.} \\
& 2_P & \text{1 in.} \quad \text{3 cm.} \\
& 3_P & \text{1·5 in.} \quad \text{4 cm.}
\end{array}
$$

A second curve may be drawn on the same paper with the distance halved.

[3]

Fig. 1. The parabola and its evolute

Geometrical Properties

In Fig. 3, SA is drawn perpendicular to AY. The curve is symmetrical about the axis AS and A is called the *vertex*. QP, qp are two positions of the variable line

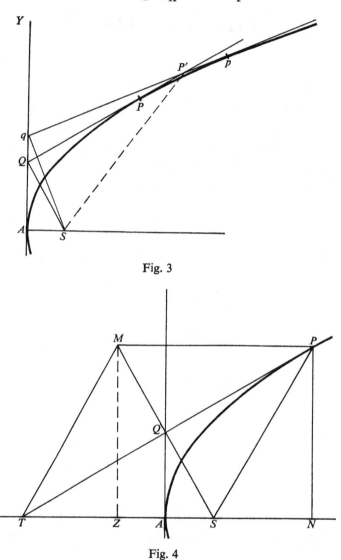

Fig. 3

Fig. 4

(i.e. two tangents to the curve), intersecting at P'. SP' is joined. Then the points S, Q, q, P' are concyclic and angle AQS = angle $qP'S$. P' is not itself a point on the curve, but, the nearer together the two tangents are, the nearer to the curve will it be. Now imagine that q moves closer to Q. P' will move towards P and angle $qP'S$ will become angle QPS. Thus, in the limit, angle AQS = angle QPS. The point P is shown again in Fig. 4.

Focus and Directrix Property

It is seen from Fig. 4 that triangles SAQ, SQP are similar; hence

$$\text{angle } ASQ = \text{angle } QSP.$$

If PQ is produced to meet the axis at T, triangles SQT, SQP are congruent. Therefore $SP = ST$. If the rhombus $PSTM$ is completed, and MZ is drawn perpendicular to ST, then $SQ = QM$ and $SA = AZ$. It follows that Z is a fixed

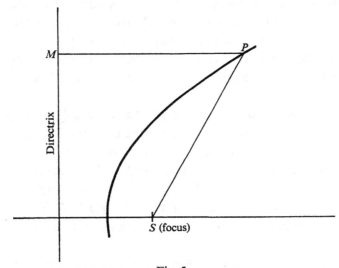

Fig. 5

point and MZ a fixed line. Moreover $SP = PM$. The parabola can thus be defined as the locus of a point P whose distance from a fixed point S (the *focus*) is equal to its distance PM from a fixed line (the *directrix*). This is shown in Fig. 5.

Cartesian Equation of the Parabola

If PN is the perpendicular from P to the axis (Fig. 4), $PN = 2QA$.

$$\therefore \quad PN^2 = 4QA^2 = 4AS.AT = 4AS.AN.$$

If AS and AQ are chosen as axes of coordinates of x and y respectively, and if P is the point (x, y), and $AS = a$, then $y^2 = 4ax$. This is the equation of the parabola.

Polar Equation of the Parabola

If, in Fig. 4, $SA = a$, $SP = r$ and angle $NSP = \theta$, then $SP = MP = ZN = ZS + SN$. Therefore $r = 2a + r\cos\theta$, and $r(1 - \cos\theta) = 2a$, or $2a/r = 1 - \cos\theta$. This is the

[5]

polar equation of the parabola, referred to S as pole and SN as initial line. The equation $2a/r = 1 + \cos\theta$ gives the same curve, turned through two right angles, since $\cos(180° + \theta) = -\cos\theta$.

Further Properties

The following may be proved as exercises:

1. If MP is produced to M', SP and PM' make equal angles with the curve (i.e. with the tangent to the curve at P).

This is the reflecting property of the parabola. If a mirror is made in the form of a paraboloid (i.e. the surface formed by rotating a parabola about its axis), rays from the focus S would be reflected into rays parallel to the axis. A searchlight beam is produced in this way. For the same reason rays coming in, parallel to the axis, would be focused at S. This is the way in which a reflecting telescope produces an accurate image of a star, free from spherical or chromatic aberration.

2. If PG (the *normal*) is drawn through P at right angles to the tangent PT, meeting the axis at G, $NG = 2a$ and is therefore constant. This gives a convenient method for drawing the normal at any point of the curve.

3. If PT cuts MZ at Y, PSY is a right angle. If PS is produced to meet the curve again at R, the tangent at R passes through Y. Angle SPY is equal to half angle RST, and angle SRY is equal to half angle PST; hence RYP is a right angle. Thus tangents at the ends of a *focal chord* meet at right angles on the directrix.

** 4. If a number of parallel chords of a parabola are drawn, their mid-points lie on a straight line parallel to the axis. This line is called a *diameter* of the parabola. The tangent at the point where it meets the curve is parallel to the chords and the tangents at the ends of any one of the chords meet on the diameter produced.

Hint: Let PP' be one of the chords and let PM and $P'M'$ be the perpendiculars from P and P' to the directrix. If K is the mid-point of MM', KS will be the radical axis of the two circles whose centres are P and P' and whose radii are PS and $P'S$. KS will thus be at right angles to the parallel chords and K will be a fixed point.

5. The two tangents from any point to a parabola subtend equal angles at the focus.

(*Hint:* If the tangents at P and P' meet at R, cutting the tangent at the vertex at Q and Q' respectively, the points S, Q, R, Q' are concyclic. Hence angle SRQ' = angle SQQ' = angle SPQ.)

[6]

Further Drawing Exercises

1. Draw any two lines and mark on each a series of points at equal intervals. (The intervals on the second line need not be equal to those on the first.) Call the points on the first line A_1, A_2, A_3, etc., and those on the second line B_1, B_2, B_3, etc. Join $A_1 B_1$, $A_2 B_2$, $A_3 B_3$, etc. The envelope of these lines will be a parabola.

2. Use the above method (i) to draw a parabola to touch four given lines; (ii) to draw a parabola to touch two given lines at given points.

(*Hints*: (i) Take two of the given lines as the two fixed lines; let the other given lines cut them at $A_1 B_1$ and $A_n B_n$. (ii) Take the two given lines as the two fixed lines; take the given points as A_1 and B_n, the intersection of the given lines being B_1 and also A_n.)

3. Draw normals at a large number of points on a parabola. (Use the property $NG = 2a$. It is convenient to mark the distance $2a$ along one edge of a set square, measuring from the right-angled corner.) The envelope of the normals drawn to any curve is called the *evolute* of that curve. The evolute of the parabola is a curve called the *semi-cubical parabola* (Fig. 1). It will be seen that from any point inside the evolute three normals can be drawn to the parabola, but from any point outside it only one.

4. Draw a circle cutting a parabola in four points and verify that the chords joining them in pairs are equally inclined to the axis.

5. Draw a circle cutting a parabola at the vertex and three other points. Verify that the normals at these three points are concurrent.

6. Given a parabola and two normals, use the last two results to draw a third normal concurrent with the first two.

7. Verify that the circumcircle of the triangle formed by three tangents to a parabola passes through the focus; and that the orthocentre of the same triangle lies on the directrix. (These properties are related to the Simson Line properties of the triangle.)

8. Draw several parabolas with the same vertex and axis, varying the position of S. Then draw a number of lines radiating from the vertex. This will illustrate the fact that all parabolas are geometrically similar: they have the same shape and vary only in size.

9. The semi-cubical parabola may be drawn as a locus as follows: Draw a fixed line AB and mark a fixed point O, not on the line. From O draw any pair of lines OL and OM, at right angles to each other, OL cutting the fixed line at Q. From Q draw QR perpendicular to AB, cutting OM at R. From R draw RP perpendicular to QR, cutting LO produced at P. Then P is a point of the locus.

Suitable dimensions. The fixed line should be near and parallel to a long edge of the paper. O may be about 2 in., or 6 cm., away from it. If graph paper is used points can be plotted very quickly by placing a ruler to represent OL and a set square with two of its sides representing OL and OM.

[7]

The Parabola: Summary

**Many of the following properties can be conveniently proved by the methods of the calculus and coordinate geometry, starting from one or other of the equations listed.

1. The Cartesian equation (origin at vertex) is $y^2 = 4ax$.
2. All parabolas are similar in shape, the constant a determining the size.
3. The polar equation (pole at focus) is $l/r = 1 - \cos\theta$, where $l = 2a$.
4. The pedal equation is $p^2 = ar$.
5. Parametric equations are $x = at^2$, $y = 2at$.
6. $\psi = 180° - \phi$, $= \frac{1}{2}\theta$, $= \cot^{-1}t$.
7. $\rho = -2a(t^2+1)^{\frac{3}{2}}$, $= -(y^2+4a^2)^{\frac{3}{2}}/4a^2$.
8. The centre of curvature is $(2a+3at^2, -2at^3)$, and the evolute is the semi-cubic parabola $27ay^2 = 4(x-2a)^3$.
9. The area bounded by the curve, the x-axis and the ordinate is $\frac{2}{3}xy$ (i.e. two-thirds of the rectangle having the same base and height).
10. $s = a[t\sqrt{(1+t^2)}+\log_e\{t+\sqrt{(1+t^2)}\}]$.
11. The parabola is the section of a right circular cone by a plane making with the axis of the cone an angle equal to the semi-vertical angle.
12. It is the negative pedal of a straight line.
13. It is the locus of a point in a plane whose distance from a fixed point (the focus) is equal to its distance from a fixed line (the directrix).
14. It is the form assumed by a hanging chain under a uniform horizontal distribution of load (cf. the catenary, p. 119).

In nos. 15–20, the notation of Fig. 4 is used. S is the focus and ZM the directrix. $SA = AZ = a$. Let PT cut MZ at R and let PG be the normal at P, meeting the axis at G.

15. SQP is a right angle.
16. PQ bisects angle SPM.
17. $SP = PM = ST = SG$.
18. $NG = 2a$.
19. PSR is a right angle.
20. If PSP' is a focal chord, the tangents at P and P' meet at right angles on the directrix, at R.
21. From a given point outside the curve two tangents can be drawn and they subtend equal angles at the focus.
22. Three tangents to a parabola form a triangle whose orthocentre lies on the directrix and whose circumcircle passes through the focus.
23. From a given point inside the evolute three normals can be drawn to the

parabola and their feet lie on a circle through the vertex; but from a point outside the evolute only one normal can be drawn.

24. The mid-points of a system of parallel chords lie on a straight line parallel to the axis. This line is called a *diameter*. The tangent at the point where it meets the curve is parallel to the chords; and the tangents at the ends of any one of the chords meet on the diameter produced.

The parabola was first studied by the Greeks as one of the sections of a cone. The earliest writer to show knowledge of these conic sections was Menaechmus (fourth century B.C.), a pupil of Plato and Eudoxus. He solved the problem of the duplication of the cube by drawing two parabolas (or, alternatively, a parabola and a hyperbola). This problem, to find the side of a cube which would have

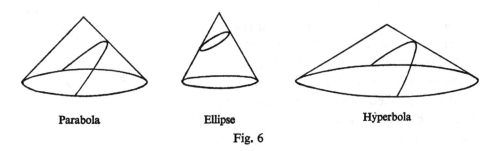

Parabola   Ellipse   Hyperbola

Fig. 6

double the volume of a given cube (in our notation, to solve the equation $x^3 = 2$ by a geometrical method), had been reduced to that of finding two geometric means between two given quantities, i.e. given a and b, find x and y such that $a/x = x/y = y/b$. The problem is insoluble by ruler-and-compass constructions, but Menaechmus solved it by finding the intersection of the parabolas $x^2 = ay$ and $y^2 = bx$. So Menaechmus evidently had some knowledge of these curves. He called the parabola a 'section of a right-angled cone', the ellipse a 'section of an acute-angled cone' and the hyperbola a 'section of an obtuse-angled cone'. This indicates that he had obtained the three curves as sections of three different right circular cones, the section being always at right angles to a generating line of the cone (Fig. 6).

Euclid wrote four books on conic sections, but they have been lost, perhaps because they were quickly superseded by the work of Apollonius (third century B.C.), 'the great geometer'. It was Apollonius who named the three curves; moreover he obtained all three from the same cone, by taking sections at different inclinations.

The origin of the names is as follows: In Fig. 7, V is the mid-point of a chord

QQ' and PV is a *diameter* (i.e. a straight line through the mid-points of a system of parallel chords). Apollonius drew a straight line PL, in the plane of the section, at right angles to PV, its length depending on the position of the section in relation to the cone. He then proved that, for the parabola, $QV^2 = PL.PV$; equivalent, in modern notation, to $y^2 = 4ax$, where x and y are oblique coordinates. The name *parabola* signifies 'equality', 'an exact comparison'.† The corresponding properties for the other two curves were $y^2 = 4ax - px^2$ and $y^2 = 4ax + px^2$; so he called one the *ellipse* ('falling short') and the other the *hyperbola* ('throwing beyond'). These names may be compared with the corresponding literary terms, *parable*, *ellipsis* and *hyperbole*.

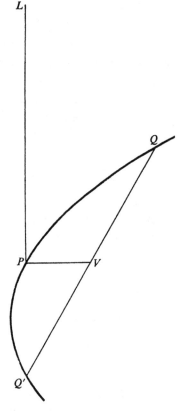

Fig. 7

Apollonius did not give the focus-directrix properties of the curves. These were first treated by Pappus of Alexandria (about A.D. 300).

The history of the conic sections begins again in the seventeenth century. The invention of coordinate geometry by Descartes put them in an altogether new light as curves of the second degree. His work on them, however, was incomplete and deliberately obscure. Wallis was the first to treat them systematically in this manner 'considered as plain Figures, exempted out of the Cone'.‡

A few years earlier, the young Pascal had treated them as projections of the circle, foreshadowing the projective geometry which was to develop 200 years later. About the same time, too, Galileo showed that the path of a projectile thrown obliquely was parabolic, a fundamental result in the science of ballistics.

The reflecting telescope was suggested by James Gregory in 1663 and the first one was made by Newton in 1668. Parallel rays are brought to a focus by a parabolic mirror, the focal length in large telescopes being anything up to 40 ft. The paraboloid form is also used in reflectors for searchlights and for radar receivers.

---

† 'Equality' as shown by 'application'; $\pi\alpha\rho\alpha\beta\acute{\alpha}\lambda\lambda\omega$ had long ago taken on the derived meaning of *comparo*.

‡ So described in *Phil. Trans. R. Soc.* (1695). Wallis's work was published in 1655.

The cables of a suspension bridge take the form of parabolas; but, as with the telescope mirrors, they show a relatively small arc, the height of the piers being usually about a tenth of the span.

The Semi-cubical Parabola: Summary

** 1. Parametric equations are $x = 3at^2$, $y = 2at^3$.

2. The corresponding Cartesian equation is $27ay^2 = 4x^3$.

3. $\tan\psi = t$.

4. The polar equation is $r = c\sin^2\theta\sec^3\theta$.

5. $s = 2a\{(1+t^2)^{\frac{3}{2}} - 1\}$.

6. The intrinsic equation is $s = 2a(\sec^3\psi - 1)$.

7. $\rho = 6at(1+t^2)^{\frac{3}{2}}$, $= 6a\sec^3\psi\tan\psi$.

8. The centre of curvature is at $(-3at^2 - 6at^4, 8at^3 + 6at)$.

9. The area bounded by the curve, the x-axis and the ordinate is $\frac{2}{5}xy$ (i.e. two-fifths of the rectangle having the same base and height).

The semi-cubic parabola became famous in its own right because it was the first algebraic curve to be rectified. William Neile devised a method (1659) for finding the length of an arc of this curve, and, in an appendix to the Latin edition of Descartes' Geometry, Heuraet used it as the most convenient example for his more general construction. Before this only transcendental curves (as, for example, the cycloid or the logarithmic spiral) had been rectified. Indeed the simplest curves seemed to defy all attempts at rectification: thus the common parabola led to logarithmic functions, while the ellipse required new functions altogether. But the discovery of Neile and Heuraet was based on the algebraic character of the 'semi-cubic' curve, not on its property as an evolute of the common parabola.

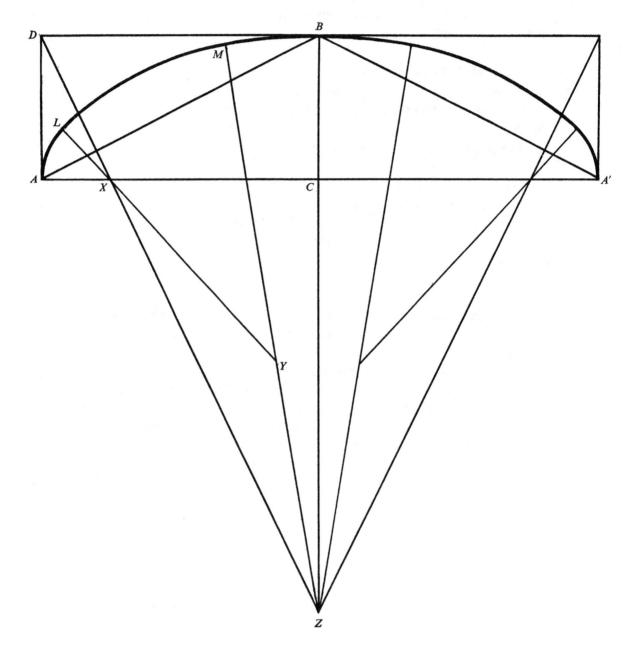

Fig. 8. A five-centred arch

2

THE ELLIPSE

To Draw an Ellipse

Draw a circle, centre C, and a diameter ACA' (Fig. 9). Mark any point S on AA'. Using a set square, draw from any point Q on the circle a chord QR at right angles to SQ. Repeat for numerous positions of Q, keeping S fixed. The envelope of QR will be an ellipse.

Suitable Dimensions

Make the circle as large as possible and make CS at least $\frac{3}{5}CA$.

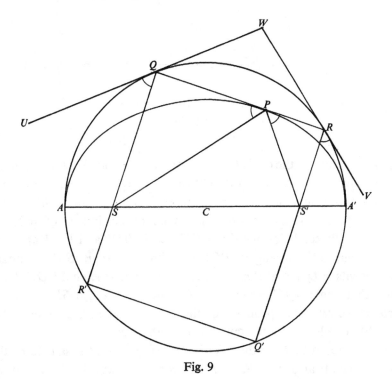

Fig. 9

Eccentricity

AA' is called the *major axis* of the ellipse and a chord BCB', drawn through C at right angles to AA' is the *minor axis*. The nearer S is to A, the narrower will be

[13]

the ellipse; i.e. the shorter will be the minor compared with the major axis. Thus ellipses can vary in shape as well as in size (unlike parabolas, which are all similar in shape: see p. 7, Ex. 8). The variation in shape is measured by the ratio $CS:CA$, which is called the *eccentricity*. It is less than 1. The length of CA (the *major semi-axis*) is usually denoted by a and the eccentricity by e. Thus $CA = a$ and $CS = ae$.

Geometrical Properties

If QS produced meets the circle again at R', RR' is a diameter; and if the rectangle $R'QRQ'$ is completed, Q' will lie on the circle. If RQ' cuts AA' at S', it is seen by the symmetry of the figure that $CS' = CS$, and that S' is thus a fixed point. S and S' are called the *foci* of the ellipse.

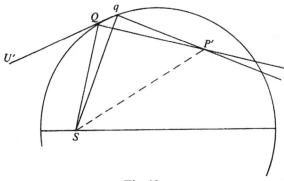

Fig. 10

Let Q, q be points on the circle (Fig. 10), and let the chord qQ be produced to U'. If the lines drawn through Q and q at right angles to SQ and Sq respectively meet at P', then the points S, Q, q, P' are concyclic, and angle $SQU' =$ angle $SP'q$. If now q approaches Q, QU' becomes in the limit the tangent to the circle at Q, and P' becomes a point on the ellipse. It follows that, in Fig. 9, if UQW and VRW are tangents to the circle, angle $SQU =$ angle SPQ; and similarly angle $S'RV =$ angle $S'PR$. But the tangents WQ and WR make equal angles with the chord QR; from which it follows that angle $SQU =$ angle $S'RV$, and hence angle $SPQ =$ angle $S'PR$. This proves the reflecting property of the ellipse, that rays emanating from one focus would be reflected by the curve to the other focus.

If SQ is produced to H so that $SQ = QH$ (Fig. 11), H is said to be the *image* of S in QR. The triangles PQS, PQH are congruent and HPS' is a straight line. Moreover, $SP + S'P = S'H$, and, since $QQ'S'H$ is a parallelogram, it follows that $SP + S'P = QQ' = AA' = 2a$. Thus the sum of the focal distances of a point on the ellipse is constant. (This explains the 'string method' of drawing an ellipse.)

[14]

Further Properties

The following may be proved as exercises:

1. The product $SQ.S'R$ is constant.

2. If the normal at P meets the major axis at G, $SG = e.SP$. (*Hint:* Use the theorem that the bisector of an angle of a triangle divides the opposite side in the ratio of the other two.)

3. If the minor semi-axis CB (Fig. 13) is of length b, use triangle CSB to prove that $b^2 = a^2(1-e^2)$.

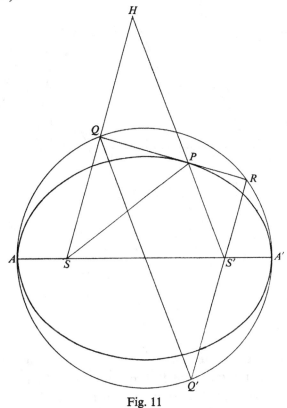

Fig. 11

4. The product $SQ.S'R = b^2$.

* 5. If $SP = r$, $S'P = r'$, $SQ = p$ and $S'R = p'$, prove that $b^2/p^2 = 2a/r - 1$. (*Hint:* Use similar triangles SPQ, $S'PR$, as well as the connections between r, r', p, p' already proved.) This result is called the *pedal equation of the ellipse*, and is useful in dealing with planetary orbits (see p. 22).

Focus and Directrix Property

* Complete the rectangle $SQPK$ (Fig. 12). Draw PN perpendicular to AA' and let CN be x. Then S, Q, P, N, K lie on a circle, and, since angle UQS = angle QPS,

[15]

QU is a tangent to that circle, as well as to the circle on AA' as diameter. Therefore the diameter QK passes through C. By intersecting chords,

$$CK.CQ = CS.CN,$$

i.e. $CK.a = ae.x,$

$$\therefore \quad CK = ex \quad \text{and} \quad SP = QK = a+ex.$$

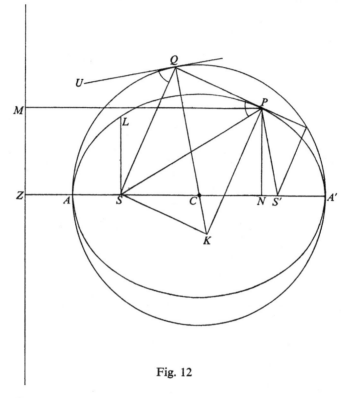

Fig. 12

This last result may be written $SP = e(a/e+x)$, from which it appears that, if a line is drawn at right angles to AA' through a point Z on CA produced, where $CZ = a/e$, and PM is drawn perpendicular to it, then $SP = e.PM$. The ellipse could thus be defined as the locus of a point whose distance from a fixed point is a fixed fraction of its distance from a fixed straight line.

Polar Equation of the Ellipse

* If $SP = r$ and angle $A'SP = \theta$,

$$r = SP = e.MP = e.(ZC-SC+SN)$$
$$= e.(a/e-ae+r\cos\theta)$$
$$= a(1-e^2)+re\cos\theta.$$
$$\therefore \quad r(1-e\cos\theta) = a(1-e^2).$$

[16]

By putting θ equal to 90°, we obtain $SL = a(1-e^2)$, where SL is the *semi-latus-rectum*, or half-width of the curve at S, measured at right angles to the major axis. SL is denoted by l, and the polar equation of the ellipse may then be written $r(1-e\cos\theta) = l$, or $l/r = 1-e\cos\theta$.

The equation $l/r = 1+e\cos\theta$ represents the same curve turned about S through two right angles.

Cartesian Equation of the Ellipse

* In Fig. 13, $SP = a+ex$ and $SN = ae+x$.

$$\therefore \quad PN^2 = SP^2 - SN^2 = (a+ex)^2 - (ae+x)^2$$
$$= a^2(1-e^2) - x^2(1-e^2).$$

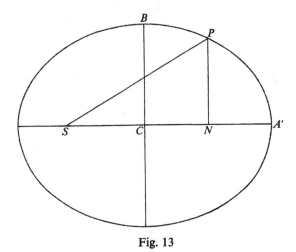

Fig. 13

If $PN = y$ so that (x,y) are the coordinates of P referred to rectangular axes CA' and CB, then

$$\frac{y^2}{a^2(1-e^2)} = 1 - \frac{x^2}{a^2} \quad \text{or} \quad \frac{x^2}{a^2} + \frac{y^2}{b^2} = 1.$$

The Ellipse and its Auxiliary Circle

* The circle on AA' as diameter is called the *auxiliary circle* of the ellipse. If (X, Y) are coordinates of a point P' on this circle, then $X^2 + Y^2 = a^2$, which may be written

$$\frac{X^2}{a^2} + \frac{Y^2}{a^2} = 1.$$

If $X = x$, as in Fig. 14, it is evident by comparison with the equation of the ellipse that $y/b = Y/a$. Hence

$$\frac{PN}{P'N} = \frac{y}{Y} = \frac{b}{a}.$$

2

[17]

LC

The ellipse might thus be drawn by reducing all the ordinates of the circle in the same ratio.

Area of the Ellipse

This can be found by comparing the ellipse with its auxiliary circle. In Fig. 14, $PN/P'N = b/a$ and, if the ellipse is divided into narrow strips by a series of ordinates PN, the area of each strip is approximately b/a times that of the corresponding part of the circle. Thus the area of the whole ellipse is $\pi a^2 \times b/a$, i.e. πab.

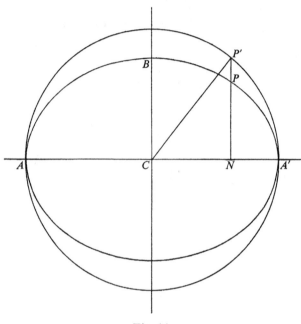

Fig. 14

Length of the Ellipse

There is no exact formula for the length of the ellipse in terms of ordinary functions. This led to the invention of *elliptic functions*, in terms of which this and other previously intractable problems could be solved.

An approximate formula given by Ramanujan in 1914 is

$$\pi[3(a+b) - \sqrt{\{(a+3b)(3a+b)\}}].$$

Further Exercises

*   1. Prove that $S'P = a - ex$. If a line is drawn at right angles to AA' through a point Z' on CA' produced, where $CZ' = a/e$, and PM' is drawn perpendicular to it, prove that $S'P = e.PM'$. (S is thus a second focus, with $Z'M$ the corresponding directrix.)

[18]

2. State the locus of the feet of the perpendiculars drawn from the foci of an ellipse to a variable tangent. (Such a locus is called a *pedal* curve; this one would be described as *the pedal of the ellipse with respect to its foci*.)

3. If, in Fig. 14, a line *PEF* is drawn parallel to *P'C* meeting the major axis at *E* and the minor axis at *F*, prove that *PF* = *a* and *PE* = *b*. (This is the basis for the *trammel method* of drawing the ellipse. See Drawing Exercise 3, below.)

4. If the normal at *P* meets the minor axis at *G'*, prove that the points *P*, *S*, *G'* and *S'* are concyclic. (*Hint:* In the triangles *PSG'* and *PS'G'*, two sides are equal to two sides, and the angles opposite one pair of sides are equal; hence the angles opposite the other pair of equal sides are either equal or supplementary.)

Hence find a construction for drawing normals to the ellipse. See p. 20, no. 9.

Further Drawing Exercises

1. Draw an ellipse by the 'string method'. (Use an endless string and the two points of a pair of dividers.) Do this again with given major and minor axes. (*Hint:* see p. 14.)

2. Draw an ellipse having given semi-axes *a* and *b*, as follows: Draw first the major axis *AA'* (= 2*a*) and a circle on *AA'* as diameter (the auxiliary circle). Draw a series of ordinates *QN* to this diameter, *Q* being any point on the circle and *N* the foot of the perpendicular from *Q* to *AA'*. On each of these ordinates mark a point *P* such that *PN*:*QN* = *b*:*a*. Draw a freehand curve through all the points *P*. (See p. 17.)

3. *The trammel method.* It is supposed that the major and minor axes are drawn. Along the edge of a piece of paper mark points *F*, *E*, *P* (in that order) such that *PF* = *a* and *PE* = *b*. Place the paper with *F* at any point on the minor axis and *E* at a point of the major axis; then *P* will be at a point of the required ellipse. (See Ex. 3, above.)

4. Take a circular piece of paper (e.g. a filter paper) and mark a point *S* on it. Fold the paper so that the edge of the folded part passes through *S*. Flatten the paper again and repeat many times. The envelope of the creases thus formed will be an ellipse. (In Fig. 11, imagine *S'* to be the centre of the piece of paper and *S'H* the radius.)

5. *Pascal's Hexagon.* Draw an ellipse and mark on it any six points *A*, *E*, *C*, *F*, *B*, *D* (in that order). Join them in the order *ABCDEF*. Mark the intersections of *AB* and *DE*, *BC* and *EF*, *CD* and *FA*. Notice the connection between these three intersections.

If the six points are joined in any order to form a hexagon, the intersections of opposite sides (i.e. the first and fourth sides, the second and fifth, the third and sixth) will be collinear.

[19]

2-2

6. *Brianchon's Theorem.* Draw any six tangents to an ellipse. (This may be done, using a set square, by the method described on p. 13.) Let these six tangents (produced if necessary) form a hexagon *ABCDEF.* Join *AD, BE* and *CF,* and verify that they are concurrent.

7. Given five points so placed that an ellipse might be drawn through them, use Pascal's Hexagon to find other points which would lie on the same ellipse.

8. Given five lines so placed that an ellipse could be drawn touching each of them, use Brianchon's Theorem to find other lines which would touch the same ellipse.

9. *The evolute of the ellipse.* A series of normals can be drawn by the method suggested on p. 19. (With centre at any point on the minor axis draw an arc through *S* cutting the ellipse at *P* and the further part of the minor axis at *G'*. Then *PG'* is a normal to the ellipse at *P*.) The envelope of these normals is the evolute, a four-cusped curve. The two cusps on the major axis are at distances b^2/a from the nearer ends of that axis; and the two on the minor axis are at distance a^2/b from further ends of that one.

10. *To draw an approximately elliptical arch with circular arcs.* Given the major axis *ACA'* and the minor semi-axis *CB,* complete the rectangle *ACBD* and draw *DXZ* at right angles to *AB,* cutting *AC* at *X* and *BC* (produced) at *Z* (Fig. 8). *X* and *Z* are then centres of curvature for the ellipse at *A* and *B* respectively, i.e. centres of the circles that would fit most closely to the desired curve at these points. (For greater accuracy it may be noted that $AX = CB^2/CA$ and $BZ = CA^2/CB$.) Now find any point *Y* within the triangle *CXZ* such that $XY + YZ = BZ - AX$. (It is best to make *XY* and *YZ* equal or nearly equal.) With centre *X,* draw an arc from *A* to meet *YX* produced at *L*; with centre *Z,* an arc from *B* to meet *ZY* produced at *M*; and with centre *Y,* an arc from *M* to *L*. (This is possible, since $MY = BZ - YZ = AX + XY = LY$.) Repeat for the other side of the arch.

Suitable dimensions. Draw *ACA'* parallel to the top edge of the paper, with *B* near the top edge.

	CA		*CB*	
Paper: 1 p	3 in. or	8 cm.	1·5 in. or	4 cm.
2 p	3 in.	8 cm.	1 in.	2·5 cm.
3 p	5 in.	12 cm.	2 in.	5 cm.

(For the arc of radius *BZ* a piece of string with a loop at one end may be used.)

The Ellipse: Summary

** 1. The Cartesian equation (origin at centre) is $x^2/a^2 + y^2/b^2 = 1$.

2. The polar equation (pole at focus) is $l/r = 1 + e\cos\theta$, where $l = a(1 - e^2)$ and $b^2 = a^2(1 - e^2)$.

[20]

3. The pedal equation (pole at focus) is $b^2/p^2 = 2a/r - 1$.

4. Parametric equations are $x = a\cos t$, $y = b\sin t$.

5. $A = \pi ab$.

6. $\rho = (a^2\sin^2 t + b^2\cos^2 t)^{\frac{3}{2}}/ab$.

7. The evolute is $(ax)^{\frac{2}{3}} + (by)^{\frac{2}{3}} = (a^2 - b^2)^{\frac{2}{3}}$.

8. The ellipse is the section of a right circular cone by a plane making with the axis of the cone an angle greater than the semi-vertical angle.

9. It is the locus of a point in a plane whose distance from a fixed point (the focus) is e times its distance from a fixed line (the directrix), e being less than 1.

10. It is the locus of a point in a plane the sum of whose distances from two fixed points is constant.

11. It is the negative pedal of a circle with respect to a point within it.

12. It is the orthogonal projection of a circle on a plane inclined to its own plane.

In nos. 13–24, the notation of Figs. 11, 12, 13, 14 is used. $CA = a$, $CB = b$, where $b^2 = a^2(1 - e^2)$. $CN = x$.

13. $CS = CS' = ae$.

14. $CZ = a/e$.

15. $SP = e \cdot PM$.

16. $SP = a + ex$, $S'P = a - ex$; $SP + S'P = 2a$.

17. $SL = l = a(1 - e^2)$.

18. QP is the external bisector of angle SPS'.

19. SQP and $S'RP$ are right angles.

20. $SQ \cdot S'R = b^2$.

21. If PEF, drawn parallel to $P'C$, meets CA' at E and BC produced at F, $PF = a$ and $PE = b$.

22. If the normal at P meets CA at G, $SG = e \cdot SP$.

23. From a point outside the curve two tangents can be drawn and they subtend equal angles at a focus.

24. If two tangents are at right angles they intersect on a circle, called the *director circle*, whose centre is C and whose radius is $\sqrt{(a^2 + b^2)}$.

25. If the tangent at P cuts the directrix at R, PSR is a right angle.

26. Tangents at the ends of a focal chord meet on the corresponding directrix.

27. From a point inside the evolute, four normals can be drawn to the ellipse; but from a point outside the evolute, only two.

28. The mid-points of a system of parallel chords lie on a straight line through the centre, called a *diameter*. The tangents at the ends of the diameter are parallel to the chords; and the tangents at the ends of any one of the chords meet on the diameter produced.

For the many geometrical properties of the ellipse the reader should consult works on the conic sections.

It has already been mentioned (p. 9) that the ellipse was studied by Menaechmus and other ancient Greeks. About 2000 years later, Kepler, working at Prague with the Danish astronomer Tycho Brahe, was set to study the motion of the planet Mars. In 1602 he wrote in a letter that he believed the orbit of Mars was oval, and later he realised that it was elliptical, with the sun at one focus. (It was he who first introduced the word 'focus' for the point in question.) Kepler published his discovery in 1609. He was unable to account theoretically for the elliptical orbit or for his other laws of planetary motion and it was left to Newton to prove (c. 1680) that the elliptical orbit was a consequence of his inverse square law of gravitation and, conversely, could only occur under such a law.

** (If a planet of mass m is moving with velocity v at distance r from the sun (mass M), its potential energy, under Newton's Law, is $-\gamma Mm/r$, and hence, by conservation of energy,

$$\tfrac{1}{2}mv^2 - \frac{\gamma Mm}{r} = \text{constant.}$$

If the perpendicular distance from the sun to the tangent to the planet's path is p, the conservation of angular momentum gives

$$pv = \text{constant } (h, \text{ suppose}).$$

Combining these two equations,

$$\frac{\tfrac{1}{2}mh^2}{p^2} - \frac{\gamma Mm}{r} = \text{constant,}$$

and this is the pedal equation of an ellipse, as shown on p. 15.)

In 1639 Pascal, at the age of 16, discovered the theorem known as Pascal's Hexagon (p. 19, Ex. 5), which is true not only for the ellipse, but for other conic sections.

James Gregory's design for a reflecting telescope (1663) included a small elliptical mirror, placed just behind the focus of the main parabolic mirror. This was for reflecting the rays to the eye-piece through a hole in the main mirror.

With the gradual acceptance of the Copernican Theory and the rotation of the earth, it became natural to think that the earth might be ellipsoidal in shape. Maclaurin (1698–1746) showed that a homogeneous mass of liquid revolving uniformly under the action of gravity would take such a form. From 1743 onwards more exact measurements became available of the length of a degree of latitude in different parts of the world. These confirmed the supposition and it is now known that the earth's polar diameter is less than its equatorial diameter by about 26 miles, i.e. by 1 part in 300. This corresponds to an eccentricity of about $\frac{1}{12}$.

The eccentricities of the planetary orbits are small, the largest being those of Mercury (about $\frac{1}{5}$) and Pluto (about $\frac{1}{4}$). For Mars the eccentricity is about $\frac{1}{11}$ and for the earth about $\frac{1}{60}$. In 1705 Halley suspected that the comet which bears his name was moving in a very elongated elliptical orbit. This has been confirmed by the regular reappearance of the comet at intervals of approximately 76 years and it has been shown that the orbit has an eccentricity of approximately 0·9675. While the orbits of the planets are nearly circular, those of the majority of comets are nearly parabolic. A few comets have been observed moving in hyperbolic orbits.

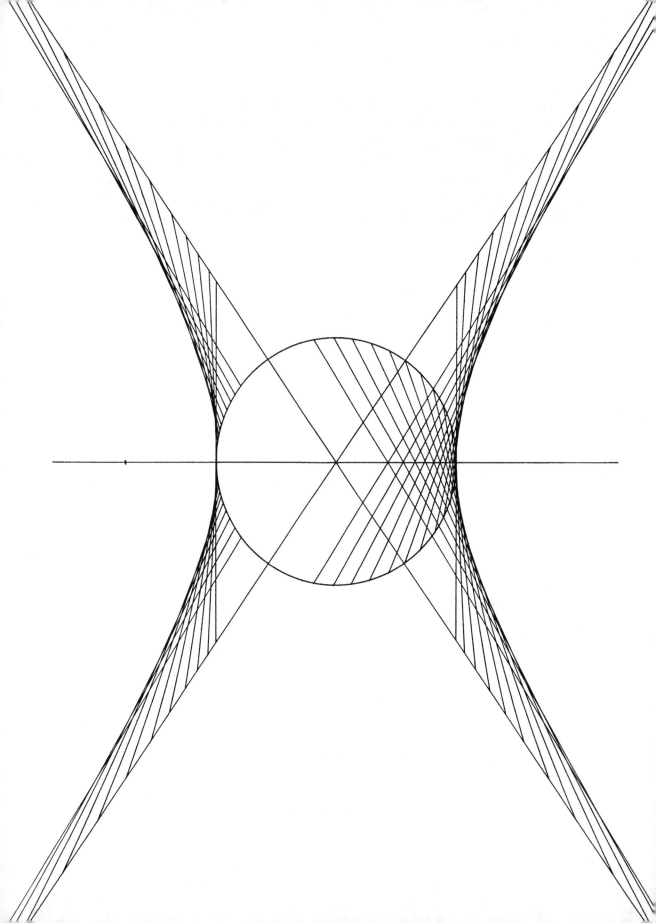

3

THE HYPERBOLA

To Draw a Hyperbola

Draw a circle, centre C, and a diameter ACA' (Fig. 16). Mark any point S on $A'A$ produced. Using a set square, draw from any point Q on the circle a chord QR at right angles to SQ. Repeat for numerous positions of Q, keeping S fixed. The envelope of RQ produced will be a hyperbola (Fig. 15).

Suitable Dimensions

The circle should be in the centre of the paper, with AA' 'horizontal'.

	CA		CS	
Paper: 1_P	1·5 in.	or 4 cm.	2·5 in.	or 6·5 cm.
1_L	1·5 in.	4 cm.	2 in.	5 cm.
2_P	2 in.	5 cm.	3·5 in.	9 cm.
2_L	2 in.	5 cm.	2·5 in.	6·5 cm.
3_P	2·5 in.	6 cm.	4·5 in.	11 cm.
3_L	2·5 in.	6 cm.	3·2 in.	8 cm.

Symmetry

If SQ meets the circle again at R', $R'CR$ will be a diameter, and the chord $R'Q'$ drawn at right angles to SR' will form with RQ a rectangle $RQR'Q'$. The envelope of this chord is part of the hyperbola and the whole figure is symmetrical about the centre C. As the hyperbola is by construction symmetrical about the axis AA' it follows that it is also symmetrical about an axis through C at right angles to AA'.

AA' is called the *transverse axis* of the hyperbola and the line through C at right angles to AA' is called the *conjugate axis*.

Eccentricity

Like the ellipse, the hyperbola varies in shape according to the position of the point S (called a *focus* of the hyperbola). The ratio $CS:CA$ is called the *eccentricity* and is denoted by e. If $CA = a$, then $CS = ae$. For the hyperbola the eccentricity is greater than 1, for the ellipse less than 1.

Asymptotes

Let q be a point on the circle near to Q, and let the lines drawn through Q and q at right angles to SQ and Sq respectively meet at P (Fig. 16). Then S, q, Q and P

[25]

Fig. 15. A hyperbola and its asymptotes

are concyclic and, if Qq is produced to W, angle SqW = angle SPQ. If now q approaches Q, P becomes in the limit a point on the hyperbola and QW becomes the tangent at Q to the circle. Angle SQW is then equal to angle SPQ.

Now let Q move along the circle from A towards T, the point of contact of the tangent from S. As it does so, the angle SQW (and hence angle SPQ) diminishes towards zero, and SQ increases. It follows that the length of QP increases indefinitely, its direction and position approximating more and more closely to those

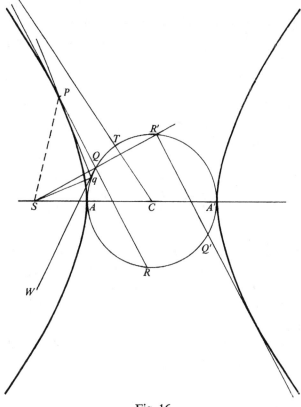

Fig. 16

of CT produced. This line CT is called an *asymptote* of the curve. It is evident by symmetry that there are two asymptotes, the curve lying close to one or other of them at all points far distant from the centre (Fig. 19).

The directions of the asymptotes are given by

$$\sec ACT = CS/CT = ae/e = e.$$

If $e = \sqrt{2}$, the asymptotes are at right angles and the curve is called a *rectangular hyperbola*.

[26]

Geometrical Properties

By symmetry RQ' meets the transverse axis at a fixed point S' (Fig. 17), and this point could be used as a second focus to construct the curve, in place of S. Let WQ and WRV be the tangents to the circle at Q and R respectively. It has been proved that angle SQW = angle SPQ, and similarly it could be shown that

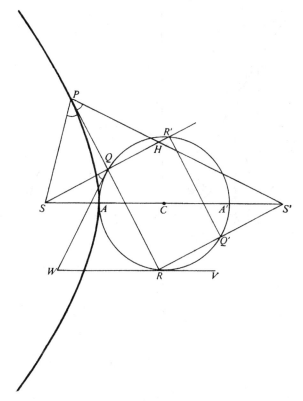

Fig. 17

angle $S'RV$ = angle $S'PR$. But angles SQW and $S'RV$ are equal, being complements of the equal angles of the isosceles triangle WRQ. Therefore angles SPQ and $S'PQ$ are equal. This proves the reflecting property of the hyperbola, that rays of light emanating from S would be reflected by the curve along lines radiating from the other focus.

If $S'P$ meets QR' at H, $QH = QS = Q'S'$. Hence $QHS'Q'$ is a parallelogram and $HS' = QQ' = 2a$. It follows that $PS' - PS = HS' = 2a$. Thus, for a point on a hyperbola, the difference of the focal distances is constant.

[27]

Focus and Directrix Property

* Complete the rectangle $SQPK$ and draw PN to meet the axis at right angles (Fig. 18). Then S, K, P, Q, N lie on a circle and, since angle SQW = angle SPQ, QW is a tangent to that circle, as well as to the circle on AA' as diameter. Therefore CQK is a straight line.

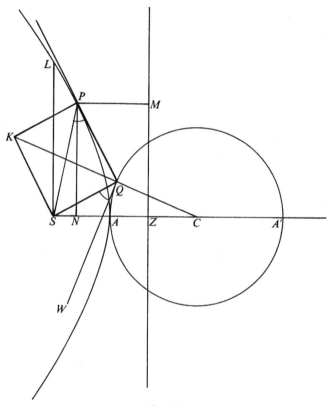

Fig. 18

By intersecting chords, $CK.CQ = CS.CN$. Let $CN = x$. Then $CK.a = ae.x$,

$$\therefore \quad CK = ex \quad \text{and} \quad SP = QK = ex - a.$$

This last result may be written $SP = e(x - a/e)$, from which it appears that, if a line is drawn at right angles to AA' from a point Z on AC, such that $CZ = a/e$, and PM is drawn perpendicular to it, then $SP = e.PM$. The hyperbola could thus be defined as the locus of a point whose distance from a fixed point is a fixed multiple (greater than 1) of its distance from a fixed straight line.

[28]

Polar Equation of the Hyperbola

* If $SP = r$ and angle $ASP = \theta$,

$$r = e.PM = e.(SC - ZC - SN)$$
$$= e(ae - a/e - r\cos\theta)$$
$$= a(e^2 - 1) - re\cos\theta,$$
$$\therefore \quad r(1 + e\cos\theta) = a(e^2 - 1).$$

By putting θ equal to $90°$ we obtain $SL = a(e^2 - 1)$, where SL is the *semi-latus-rectum*, or half-width of the curve at S, measured at right angles to the transverse axis. If SL is denoted by l, the polar equation of the hyperbola may be written

$$r(1 + e\cos\theta) = l \quad \text{or} \quad l/r = 1 + e\cos\theta.$$

It will be noted that, as θ increases from zero, r increases as long as $\cos\theta > -1/e$, increasing without limit as that value is approached. When $\cos\theta < -1/e$, r is negative and points are obtained on the further branch of the curve.

The equation $l/r = 1 - e\cos\theta$ represents the same curve turned about S through two right angles.

Cartesian Equation of the Hyperbola

* In Fig. 18,
$$SP = ex - a \quad \text{and} \quad SN = ae - x.$$
$$\therefore \quad PN^2 = SP^2 - SN^2 = (ex - a)^2 - (ae - x)^2$$
$$= x^2(e^2 - 1) - a^2(e^2 - 1).$$

If $PN = y$, so that (x, y) are coordinates of P referred to rectangular axes through C, then

$$\frac{y^2}{a^2(e^2 - 1)} = \frac{x^2}{a^2} - 1 \quad \text{or} \quad \frac{x^2}{a^2} - \frac{y^2}{b^2} = 1,$$

where $b^2 = a^2(e^2 - 1)$.

The length b may be represented by CB in Fig. 19, for it was shown above (p. 26) that $\sec ACT = e$, and therefore $CB^2/CA^2 = \tan^2 ACT = e^2 - 1$.

Rectangular Hyperbola

For the rectangular hyperbola $\tan ACT = 1$, and $e = \sqrt{2}$. Hence $CB = CA$, or $b = a$. The Cartesian equation is then $x^2 - y^2 = a^2$.

Drawing Exercises

1. *To draw a hyperbola (second method).* Given the asymptotes CX, CY, and a tangent ST, meeting them at S and T, the curve may be drawn as follows:

On SC produced mark T' so that $CT' = CT$. Draw any circle through S and T', and a number of chords *PCQ*. Mark off *CP'*, equal to *CP*, along one asymptote and *CQ'*, equal to *CQ*, along the other. Then the envelope of *P'Q'* will be the required hyperbola.

Suitable dimensions. For the asymptotes draw diagonals of the paper. The circle through *S* and *T'* should have as large a radius as possible. (It is sufficient if about half the circle is on the paper.)

2. *To draw a hyperbola* (*third method*). Given the asymptotes *CX*, *CY*, and a point *Q* on the curve, draw a line through *Q* cutting the asymptotes at *U*, *V*. On

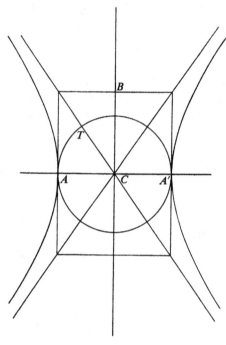

Fig. 19

this line mark a point *P* such that *PV* = *UQ* (the direction on the line being indicated by the order of the letters). The locus of such points *P* will be the hyperbola.

3. *To draw a hyperbola mechanically.* Fix the ends of two strings at the proposed positions of the foci. Tie the other ends together, arranging the lengths so that their difference is equal to the desired length of the transverse axis. Thread both strings through a small ring and place a pencil in the ring. If the ring and pencil are now moved so that the strings are kept taut the pencil will describe one branch of a hyperbola.

4. *To draw the tangent and normal at any point of a hyperbola.* Draw the internal and external bisectors of angle *SPS'*. Alternatively, suppose that in Fig. 17, the

tangent at P meets the conjugate axis at T'. Then $ST' = S'T'$. In triangles PST' and $PS'T'$ we also know that angles SPT' and $S'PT'$ are equal, and PT' is common. Therefore, since angles PST' and $PS'T'$ are not equal, they are supplementary. Therefore the points P, S, T' and S' are concyclic and, if the circle cuts the conjugate axis again at G', $T'G'$ is a diameter and PG' is the normal to the hyperbola at P. Hence the following construction:

Let the perpendicular bisector of SP cut the conjugate axis at K. With centre K and radius KS draw a circle cutting the conjugate axis at T' and G', G' being on the same side of the transverse axis as P. Then PT' is the tangent and PG' the normal.

5. *To draw the evolute.* With any point on the conjugate axis as centre draw a circle through S, cutting the curve at P and the conjugate axis at G' (on the same side of the transverse axis as P). Then PG' is a normal and the envelope of such lines is the evolute. This is a curve with two cusps on the transverse axis, pointing inwards towards the centre, the distances from the cusp-points to the centre being $(a^2+b^2)/a$.

** 6. Draw a rectangular hyperbola and mark any three points A, B and C on it. The orthocentre of triangle ABC will be found to lie on the hyperbola, and the nine-points circle of the triangle will pass through the centre of the hyperbola.

The Hyperbola: Summary

** 1. The Cartesian equation (using the transverse and conjugate axes of the curve as axes of coordinates) is $x^2/a^2 - y^2/b^2 = 1$.

2. The Cartesian equation (using the asymptotes as axes) is $xy = c^2$, where $4c^2 = a^2 + b^2$.

3. The polar equation (pole at focus) is $l/r = 1 + e\cos\theta$, where $l = a(e^2 - 1)$ and $b^2 = a^2(e^2 - 1)$.

4. The pedal equation (pole at focus) is $b^2/p^2 = 2a/r + 1$.

5. Parametric equations (using the transverse and conjugate axes as axes of coordinates) are

$$x = a\sec\theta, \quad y = b\tan\theta \quad \text{(not the } \theta \text{ of the polar equation),}$$
or $$x = a\operatorname{ch}u, \quad y = b\operatorname{sh}u \quad \text{(one branch only).}$$

6. Parametric equations (using the asymptotes as axes) are $x = ct$, $y = c/t$.

7. Using the parameter θ of no. 5 above,

$$\rho = \pm(a^2\tan^2\theta + b^2\sec^2\theta)^{\frac{3}{2}}/ab.$$

8. The evolute (using the axes of the curve as axes of coordinates) is

$$(ax)^{\frac{2}{3}} - (by)^{\frac{2}{3}} = (a^2+b^2)^{\frac{2}{3}}.$$

[31]

9. For the rectangular hyperbola, $e = \sqrt{2}$ and $b = a$. The Cartesian equation, as in (1) above, is $x^2 - y^2 = a^2$. With the asymptotes as axes, the Cartesian equation is $xy = c^2$, where $2c^2 = a^2$, and the parametric equations are $x = ct$, $y = c/t$. The polar equation, with the pole at the centre and the transverse axis as initial line, is $r^2\cos 2\theta = a^2$.

10. The hyperbola is a section of a right circular (double) cone by a plane making with the axis of the cone an angle less than the semi-vertical angle.

11. It is the locus of a point in a plane whose distance from a fixed point (the focus) is e times its distance from a fixed line (the directrix), e being greater than 1.

12. It is the locus of a point in a plane the difference of whose distances from two fixed points is constant.

13. It is the negative pedal of a circle with respect to a point outside it.

14. It is the envelope of a line the product of whose intercepts on two fixed lines, measured from their point of intersection, is constant.

In nos. 15–26, the notation of Figs. 17, 18 is used. $CA = a$ and $b^2 = a^2(e^2 - 1)$. $CN = x$.

15. $CS = CS' = ae$.

16. $CZ = a/e$.

17. $SP = e.PM$.

18. $SP = ex - a$, $S'P = ex + a$; $S'P - SP = 2a$.

19. $SL = l = a(e^2 - 1)$.

20. QP bisects angle SPS'.

21. SQP and $S'RP$ are right angles.

22. $SQ.S'R = b^2$.

23. If the normal at P meets CA produced at G, $SG = e.SP$.

24. If a straight line cuts the curve, the intercepts on it between the curve and the asymptotes are equal.

25. From a point between the two branches of the curve two tangents can be drawn and they subtend equal angles at a focus.

26. If two tangents are at right angles they intersect on a circle, called the *director circle*, whose centre is C and whose radius is $\sqrt{(a^2 - b^2)}$.

27. If the tangent at P cuts the directrix at R, PSR is a right angle.

28. Tangents at the ends of a focal chord meet on the corresponding directrix.

29. From a point between the two branches of the evolute two normals can be drawn to the hyperbola; but from a point beyond the evolute four.

30. The mid-points of a system of parallel chords lie on a straight line through the centre, called a *diameter*. The tangents at the ends of the diameter are parallel to the chords; and the tangents at the ends of any one of the chords meet on the diameter produced.

[32]

31. If a triangle is formed by joining three points on a rectangular hyperbola, its orthocentre lies on the curve and its nine-points circle passes through the centre of the curve.

For the many geometrical properties of the hyperbola the reader should consult works on the conic sections.

It was the rectangular hyperbola that Menaechmus used, and particularly its asymptote property ($xy = ab$). The lost works of Aristaeus, on *Solid Loci*, and Euclid, on *Conics*, probably dealt with the general hyperbola, but only a single branch of it. It was Apollonius who first treated the double-branched curve, obtaining it as a section of a double cone.

Cassegraine's design for a reflecting telescope (1672) included a small convex hyperbolic mirror placed just in front of the focus of the main parabolic mirror. This was for reflecting the rays to the eye-piece through a small hole in the main mirror.

A branch of a hyperbola may often be seen as the edge of the shadow cast on a wall by a circular lampshade.

The property that the difference of the focal distances is constant has given the hyperbola an important place in the theory and practice of sound-ranging and of radar navigation.

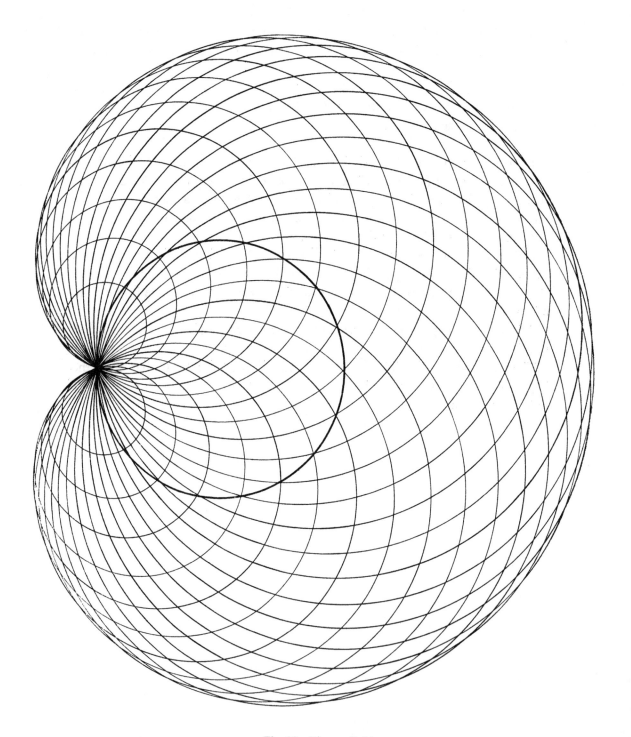

Fig. 20. The cardioid

4

THE CARDIOID

To Draw a Cardioid

Draw a circle (to be called the *base-circle*) and mark a fixed point A on it. With centre at any point Q on the circle, and radius QA, draw another circle. Repeat for a large number of positions of Q, spread evenly round the base-circle. The heart-shaped curve which all these circles touch is the *cardioid* (Fig. 20). The pointed part at A is called a *cusp*, and A is called the *cusp-point*.

Suitable dimensions (with A on the left-hand side of the base-circle):

		Radius of base-circle	Distance of centre from left-hand edge of paper
Paper:	1_P	1·3 in. or 3 cm.	3 in. or 8 cm.
	2_P	1·5 in. 4 cm.	3 in. 8 cm.
	3_P	2·0 in. 5 cm.	4·5 in. 12 cm.

It is best to begin with the point Q on the far side of the base-circle from A and to move it gradually towards A.

Measure the lengths of chords PAP' of the cardioid, and find where their midpoints lie. Draw tangents at the ends of one of these chords and notice the angle at which they intersect. (See Summary, nos. 2, 5.)

Geometrical Properties

If Q and q are the centres of two circles through A which meet again at P (Fig. 21), triangles QAq and QPq are congruent, and P is the *image* of A in the line Qq (i.e. Qq is the perpendicular bisector of AP). If q now moves close to and, in the limit, coincides with Q, Qq becomes the tangent to the base-circle at Q; and P becomes a point on the cardioid.

It thus appears that the cardioid could be drawn by constructing a series of image points of A in tangents to the base-circle; or on half-scale as the locus of the foot of the perpendicular from A to a tangent to the base-circle. A curve drawn in this way is called the *pedal* of the base-curve; thus the cardioid is said to be the pedal of a circle with respect to a point on its circumference. (See ch. 18, p. 153.)

[35]

If P is a point on the cardioid corresponding to a point Q on the base-circle, and QQ' is a diameter of the base-circle (Fig. 22), there will be another point P' of the cardioid corresponding to Q'.

The following facts may now be proved as exercises:

1. If the radius of the base-circle is a, $PP' = 4a$.

2. PQ and $P'Q'$ meet at right angles, at a point R on the base-circle.

3. The mid-point M of PP' is on the base-circle. (*Hint:* Let PP' meet the tangent at Q in T and the tangent at Q' in T'; then $MT = AT'$, by symmetry. Hence find $P'M$.)

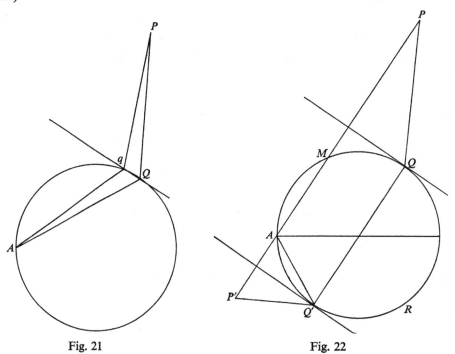

Fig. 21   Fig. 22

4. The tangents to the cardioid at P and P' are at right angles and the locus of their intersection is a circle. (*Hint:* The tangent to the cardioid at P is the same as the tangent to the circle whose centre is Q and whose radius is QP: it is therefore at right angles to QP. Consider the quadrilateral formed by PQ, $P'Q'$ and the two tangents.)

Polar Equation of the Cardioid

From Ex. 3 above, it is seen that another way to construct points on the cardioid would be to draw chords AM of the base-circle and to produce them to P and P' so that $MP = MP' = 2a$ (Fig. 23). A curve drawn in this way is called a *conchoid*;

[36]

thus the cardioid is the conchoid of a circle with respect to a point on its circumference, with the fixed distance equal to the diameter of the circle. (See ch. 14, p. 127.)

If AP makes an angle θ with the diameter AB of the base-circle, $AM = 2a\cos\theta$, and $AP = 2a + 2a\cos\theta$. If the length of AP is called r, then $r = 2a(1 + \cos\theta)$. (It should be noted that, if θ is given a value 180° more than for P, the point P' is obtained.)

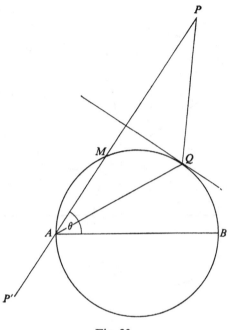

Fig. 23

The Cardioid as an Epicycloid

Imagine that, in Fig. 24, the tangent at Q to the base-circle is a mirror; it has already been seen that P is the image of A in such a mirror. The image of the base-circle would be an equal circle, passing through P and touching the base-circle at Q (Fig. 24). The arc PQ would be equal to the arc AQ, and it follows that, if this image circle were rolled round the outside of the base-circle, P would eventually arrive at A. Thus P is a point fixed on the circumference of the rolling circle, and the cardioid is the locus of a point on the circumference of a circle which rolls round the outside of an equal fixed circle. (This may be illustrated with two coins; pennies may be used, but half-crowns are better, because of the milled edges.) A curve formed in this way is called an *epicycloid*.

[37]

Double Generation

Now consider the circle on PP' as diameter (Fig. 25). As its centre is at M, it touches the base-circle at R; and as angle AMR is half angle AOR, its arc $P'R$ is equal to the arc AR of the base-circle (radius double, angle at centre half). From this it follows that, if it were rolled on the base-circle, P' would eventually arrive at A (and so would P, after another half-turn); thus the cardioid is the locus of a

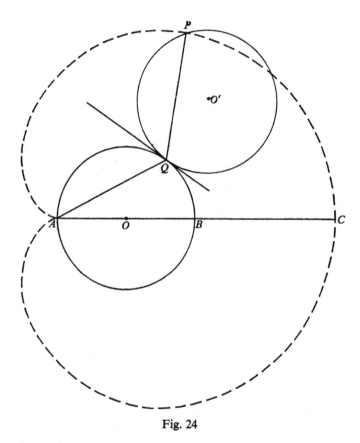

Fig. 24

point on the circumference of a circle rolling on a fixed circle of half its radius, the fixed circle being inside the rolling circle.†

Further Drawing Exercises

1. Draw the cardioid as the pedal of a circle with respect to a point on its circumference. (Use a set square.)

† Properties of the cardioid are illustrated in a film by T. J. Fletcher of the Sir John Cass College, London, E.C. 3.

2. Draw the cardioid as the conchoid of a circle with respect to a point on its circumference.

3. Draw a cardioid on squared paper and measure its area by counting squares. Compare the result with the area of the base-circle. (See Summary, no. 11.)

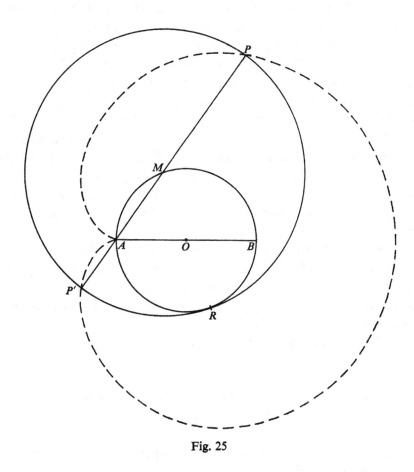

Fig. 25

4. Draw a parabola with the fixed line about $4\frac{1}{2}$ in. from the short edge of your paper and the focus S a further $\frac{1}{4}$ in. away. If P is any point on the parabola, draw SP and measure its length (r in.). On SP mark a point Q such that $SQ = 1/r$ in. Do this for many positions of SP. The locus of Q is a cardioid. This process is called *inversion*; the cardioid is said to be the *inverse* of the parabola with respect to its focus.

5. On any cardioid draw three parallel tangents; join their points of contact to the cusp-point A and notice the angles these lines make with each other. (See Summary, no. 6.)

6. Draw a chord cutting the cardioid in four points; join each of these points to A and verify that the sum of their lengths is $4a$.

* 7. Draw two unequal lines RO and RO' intersecting at R so that angle ORO' is 120°. With centres at O and O' draw circles intersecting at R. Place a 60° set square so that the sides of the 60° angle each touch one of these circles (on the inside of the angle) and mark the position of the vertex of that angle. The locus of this point, as the position of the set square varies, is a cardioid.

Suitable dimensions: With paper 1_P the radii may be 1 in. and $\frac{3}{4}$ in., or 3 cm. and 2 cm., or (with small set squares) 1 in. and $\frac{1}{2}$ in.; with larger paper, $1\frac{1}{2}$ in. and 1 in. or 5 cm. and 3 cm. The side of a small set square can be 'lengthened' by placing a ruler alongside it. The locus will be found to touch the two circles at the points where OR produced and $O'R$ produced meet them again. Between those points and the cusp, which is at R, the point of contact of one side is on that side produced, and a ruler must be placed alongside to obtain the position.

Proof: Let the sides of the set square be AB and AC, touching the circles at P and Q respectively. Draw OA' and $O'A'$ parallel to BA and CA, to meet at A'. Draw perpendiculars $A'P'$ and $A'Q'$ from A' to AB and AC, meeting them at P' and Q'. Then triangles $P'A'Q'$ and ORO' are congruent, and the circle on AA' as diameter, which passes through P' and Q', is the same size as the fixed circle $ORO'A'$. Moreover, angle $OA'R$ is equal to angle $P'AA'$, since they subtend equal chords in equal circles. Therefore $RA'A$ is a straight line, and $A'A$ is of constant length. It follows that A lies on a cardioid.

For a fuller treatment of this problem, see p. 50.

The Cardioid: Summary

** 1. The polar equation is $r = 2a(1-\cos\theta)$. (The change of sign corresponds to changing θ into $180°+\theta$. The cardioid is thus turned through 180° and is now placed as in Fig. 26.)

2. The length of any chord through the cusp-point is $4a$, and the mid-points of such chords lie on a circle.

3. $\phi = \frac{1}{2}\theta$.

4. $\psi = \frac{3}{2}\theta$.

5. The tangents at the ends of any chord through the cusp-point are at right angles.

6. There are three parallel tangents in any given direction and the values of θ at their points of contact differ by multiples of 120°.

7. The inverse of a cardioid with respect to its cusp-point is a parabola with its focus at that point.

8. The pedal equation is $r^3 = 4ap^2$.

9. The intrinsic equation is $s = 8a(1 - \cos\frac{1}{3}\psi)$, where $0 < \psi < 540°$.

10. $L = 16a$.

11. $A = 6\pi a^2$.

12. $\rho = \frac{8}{3}a|\sin\frac{1}{2}\theta|$.

13. The cardioid is the conchoid of a circle with respect to a point on its circumference, the fixed distance being equal to the diameter of the circle.

14. It is the pedal of a circle with respect to a point on its circumference.

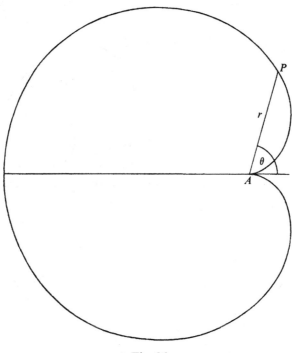

Fig. 26

15. It is the epicycloid formed by a point on the circumference of a circle which rolls with external contact on a fixed circle of the same radius; or which rolls with internal contact round a fixed circle of half the radius.

16. It is the caustic of a circle with respect to a point on its circumference, i.e. if rays emanating from that point are reflected by the circumference when they meet it again, the envelope of the reflected rays is a cardioid.

(*Hint for proof:* Suppose that, in Fig. 27, the circle whose centre is O' rolls round the outside of the fixed equal circle whose centre is O, the point P tracing a cardioid whose cusp-point is at A. Let OO' meet the fixed circle at Q and the rolling circle at T. Then P is moving at right angles to QP and PT is, therefore, a tangent to the

[41]

cardioid. Let a circle be drawn with centre O, radius OT, and let AB be produced to meet it at C. It is now only necessary to prove that a ray CT would be reflected by the circle at T along TP.)

17. From Fig. 27, parametric equations for P, with O as origin and θ replaced by t, are $\qquad x = 2a\cos t - a\cos 2t, \quad y = 2a\sin t - a\sin 2t.$

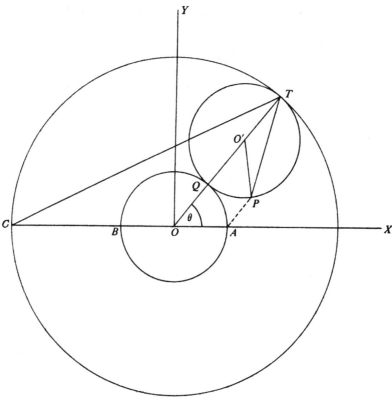

Fig. 27

(*Hint*: The gradient of PO' is $2t$. For the x-coordinate, project OO' and $O'P$ on to OX; for the y-coordinate, project them on to OY.)

For the point diametrically opposite to P on the rolling circle,

$$x = 2a\cos t + a\cos 2t, \quad y = 2a\sin t + a\sin 2t.$$

(The first pair of equations represent a cardioid orientated as in Fig. 26, the second pair as in Fig. 24.)

18. The evolute is a cardioid with base-circle having the same centre as that of the original cardioid, but one-third of the radius, the orientation of the evolute being opposite to that of the original curve.

[42]

(*Hint:* In Fig. 27, PQ is the normal. If it is produced to R, AQ and QR are equally inclined to OQ. Hence its envelope is the caustic of circle AQB with respect to A. Alternatively, with origin at the centre of the base-circle, a point of the evolute is given by

$$x = a + r\cos t - \rho \sin \psi = \tfrac{1}{3}a\cos 2t + \tfrac{2}{3}a\cos t;$$
$$y = r\sin t + \rho \cos \psi \quad = \tfrac{1}{3}a\sin 2t + \tfrac{2}{3}a\sin t.)$$

The name *cardioid* ('heart-shaped') was first used by de Castillon in the *Philosophical Transactions of the Royal Society* of 1741, but he was not the first to consider the curve: its length, for instance, had been found by La Hire in 1708.

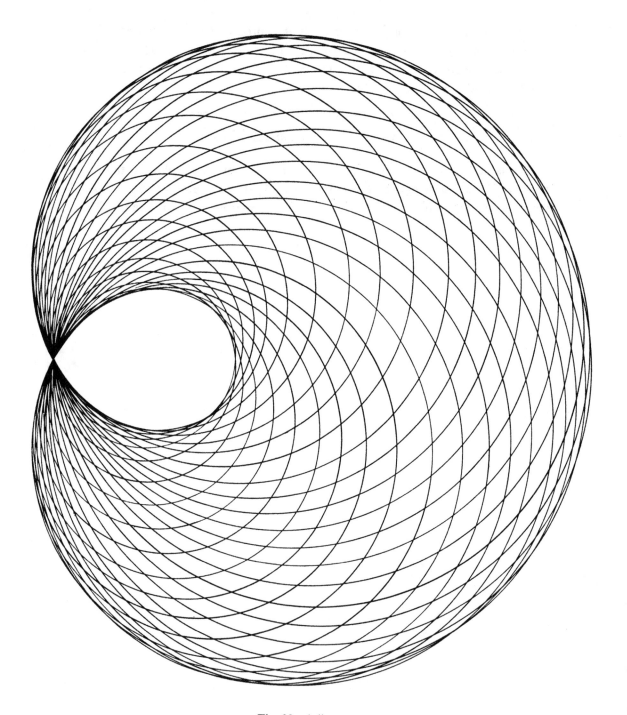

Fig. 28. A limaçon

5

THE LIMAÇON

To Draw a Limaçon

Draw a circle and a diameter $AB (= 2a)$. If Q is the middle point of one edge of your ruler, place the ruler with Q on the circle, the same edge passing through A

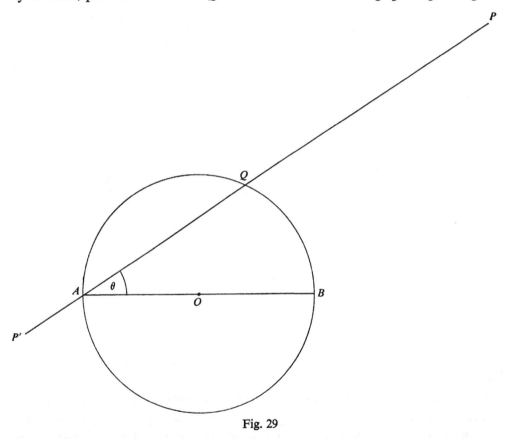

Fig. 29

(Fig. 29). Mark two points P, P' at a fixed distance k on either side of Q, k being greater than $2a$. Repeat this many times. (It is best to begin with Q at B and to move Q gradually round the circle.) Draw a freehand curve through the marked points. This curve is the *limaçon*.

[45]

Draw another limaçon with k less than $2a$.

Suitable dimensions (with A on the left-hand side of the circle):

	Radius (a)		Distance of centre from left-hand edge of paper		k (first value)		k (second value)	
Paper: 1_P	1 in.	2·5 cm.	2·5 in.	7 cm.	3 in.	7·5 cm.	1 in.	2·5 cm.
2_P	1 in.	2·5 cm.	3 in.	8 cm.	3 in.	7·5 cm.	1 in.	2·5 cm.
3_P	1·5 in.	4 cm.	4 in.	10 cm.	4·5 in.	12 cm.	1·5 in.	4 cm.

Special Cases

When $k = 2a$, the curve is a *Cardioid* (see p. 36). When $k = a$, the curve is the *Trisectrix* (Fig. 30).†

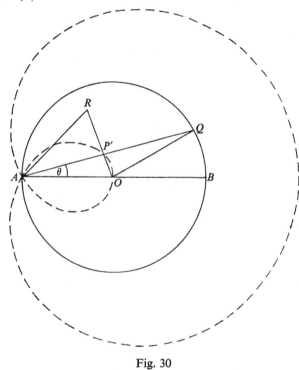

Fig. 30

The Trisectrix

This curve can be used for trisecting an angle, as follows: Make an angle BAR equal to the given angle, with AR equal to the radius a (Fig. 30). Join R to O, the centre of the circle, cutting the inner loop of the trisectrix at P'. Join AP'. Then angle $BAP' = \frac{1}{3}$ of angle BAR.

Proof: Produce AP' to meet the circle at Q (Fig. 30); then $P'Q = a$, since P' is

† Not the same as the Trisectrix of Maclaurin, $r = a\sec(\frac{1}{3}\theta)$, or the Trisectrix of Catalan, $r = a\sec^3(\frac{1}{3}\theta)$. See pp. 154, 157.

[46]

one of the points of the trisectrix arising from Q on the circle. Join OQ, and let angle OAQ be θ. Then

$$OA = OQ, \quad \therefore \text{ angle } Q = \theta \text{ and angle } AOQ = 180° - 2\theta;$$
$$QO = QP', \quad \therefore \text{ angle } P'OQ = 90° - \tfrac{1}{2}\theta.$$

Hence angle $AOR = 90° - \tfrac{3}{2}\theta$; but

$$AO = AR, \quad \therefore \text{ angle } OAR = 3\theta.$$

The same curve can be used for drawing angles of 36° and 72°, and hence for drawing a regular pentagon. Let the perpendicular bisector of AO meet the outer branch at P_1. If AP_1 cuts the base-circle at Q_1, $Q_1P_1 = Q_1O$. Then, if angle $OAP_1 = \theta$,

$$P_1A = P_1O, \qquad \therefore \quad \text{angle } AP_1O = 180° - 2\theta;$$
$$Q_1P_1 = Q_1O, \qquad \therefore \quad \text{angle } AQ_1O = 360° - 4\theta;$$
$$OA = OQ_1, \qquad \therefore \quad \theta = 360° - 4\theta,$$
$$\therefore \quad \theta = 72°.$$

In a similar manner it may be proved that, if the perpendicular bisector meets the inner branch at P_2, angle $OAP_2 = 36°$.

Polar Equation of the Limaçon

If angle BAQ (Fig. 29) $= \theta$, and $AP = r$, then $AQ = 2a\cos\theta$, and $r = 2a\cos\theta + k$. This is the polar equation of the limaçon. It is not necessary to give a separate equation for P', because that point is obtained when the value of θ is increased by 180°.

For the cardioid, $\qquad\qquad r = 2a(\cos\theta + 1);$

for the trisectrix, $\qquad\qquad r = a(2\cos\theta + 1).$

The Limaçon as a Pedal Curve

Draw a circle of radius k and mark a fixed point at a distance $2a$ from the centre. On any tangent to the circle mark the foot of the perpendicular from A. The locus of such points is a limaçon.

Suitable dimensions are as for the first method.

Proof: In Fig. 31, TP and $T'P'$ are parallel tangents to the circle whose centre is B; P and P' are the feet of the perpendiculars from A, and BQ is parallel to the two tangents. Then AQ is a chord of a circle on AB as diameter and $QP = QP' = k$. Therefore the locus of P and P' is a limaçon.

[47]

The Limaçon as an Epitrochoid

* If a circle rolls round the outside of a fixed circle, the locus of a point carried by it, but not on its circumference, is called an *epitrochoid*.

In Fig. 32, A, O, B, Q, P are as before. A circle S_1 is drawn with centre O and radius $\tfrac{1}{2}k$, cutting AB at U and V. Another circle S_2, of the same radius, is drawn with centre O', where OO' is equal and parallel to QP. These two circles touch at T.

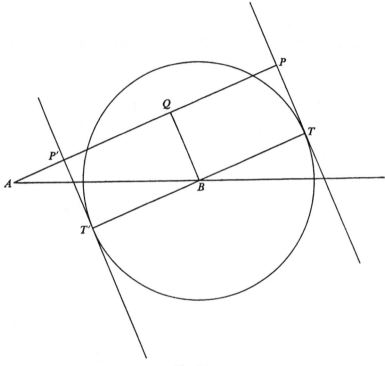

Fig. 31

Then $PQOO'$ is a parallelogram and $O'P = OQ = OA = a$. If PO' is produced to meet the circle S_2 at V',

$$\text{angle } TO'V' = \text{angle } P = \text{angle } AQO = \text{angle } QAO = \text{angle } TOV.$$

It follows that arc $TV' = $ arc TV, and circle S_2 could be rolled round circle S_1 until V' coincides with V. Thus V' is a point fixed on the circumference of the rolling circle, and the locus of P is that of a point fixed to the circle S_2 as it rolls round circle S_1.

It may further be noted that, in Fig. 32, the circle S_2 is turning momentarily about the point T (the *instantaneous centre*). Thus P is moving at right angles to TP. Moreover, $TP = TA$. This explains the following method for drawing the limaçon.

[48]

Envelope Method for Drawing the Limaçon

Let A be a fixed point at a distance a from the centre of a base-circle of radius $\frac{1}{2}k$. With any point T on the base-circle as centre, and radius TA, draw a circle. The envelope of such circles will be the limaçon (Fig. 28).

Suitable Dimensions

The values of a and k should be as given for the first method, and the centre of the circle should be placed in the position stated. The point A should be on the left of the centre. It is recommended that complete circles should be drawn, even when $k < 2a$.

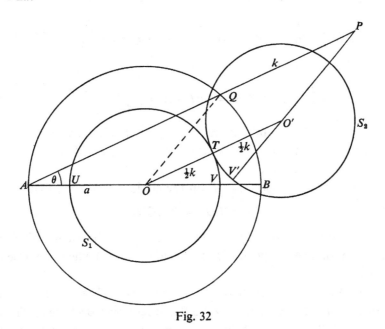

Fig. 32

Dürer's Method of Drawing the Limaçon

A circle is drawn and the circumference divided into twelve parts, these being numbered like the hour divisions of a clock-face. Radii are drawn to these twelve points. From the ends of radii 1, 2, 3, ..., lines are drawn, all the same length, parallel to the radii 2, 4, 6, ..., respectively. The curve is then drawn through the ends of these lines. (In Dürer's diagram, the first 'hour' is subdivided into ten parts, presumably as an indication that more points may be plotted if desired. This diagram is reproduced in J. L. Coolidge, *The Mathematics of Great Amateurs*, p. 67.)

A Locus Problem

If a triangle moves so that two of its sides touch respectively two fixed circles, the locus of the vertex at which those sides meet consists of four limaçons.

* *Proof:* Let the centres be O and O' and let one of their intersections be K. Let the sides AB and AC of the triangle touch the circles at P and Q respectively. Draw OA' and $O'A'$ parallel to BA and CA, to meet at A'. Then A' moves on a fixed circle through O and O'. Let the diameter of this circle be d, and let AA' meet the circle again at R. Let $A'P'$ and $A'Q'$ be the perpendiculars from A' to AB and AC respectively.

Then $OR = d \sin OA'R$ and $RO' = d \sin RA'O'$.

$$\therefore \quad \frac{OR}{RO'} = \frac{\sin OA'R}{\sin RA'O'} = \frac{\sin P'AA'}{\sin A'AQ'} = \frac{P'A'}{A'Q'} = \frac{PO}{O'Q} = \text{constant.}$$

Therefore R is a fixed point.

Moreover $\qquad \dfrac{OR}{d} = \sin OA'R = \sin P'AA' = \dfrac{P'A'}{AA'} = \dfrac{PO}{AA'}.$

Therefore AA' is of constant length, and the locus of A is a limaçon.

The limaçon has a loop if $AA' < d$, or $OP < OR$, i.e. if R is outside the two circles. This occurs if angle $OKO' > 180° -$ angle A.

There is also another position of R on the circle $A'OO'$ such that

$$OR/RO' = PO/O'Q.$$

These two positions correspond to the two distinct cases (i) when both circles touch the sides of the triangle on the inside (or outside), and (ii) when one touches on the inside and one on the outside.

Moreover, the figure is symmetrical about OO'. Hence the complete locus consists of four limaçons. Let OKO' be θ and let angle BAC be α. Then, if θ lies between α and $(180° - \alpha)$, two of these limaçons have loops; if θ is greater than each of those values, all four have loops; and if θ is less than each of those values, none of them have loops. If θ is equal to α, or $(180° - \alpha)$, the corresponding limaçon is a cardioid; and, if $\theta = \alpha = 90°$, the four limaçons reduce to two cardioids.

The Limaçon: Summary

** 1. The polar equation is $r = k + 2a\cos\theta$, or (with the direction of the initial line reversed) $r = k - 2a\cos\theta$.

2. Parametric equations (with O as origin and OB as axis of x) are

$$x = k\cos t + a\cos 2t, \quad y = k\sin t + a\sin 2t.$$

[50]

3. If $k = 2a$, the curve is the cardioid; if $k = a$, it is the trisectrix.

4. If $k > 2a$, $A = (2a^2 + k^2)\,\pi$;

    if $k = 2a$, $A = 6\pi a^2$;

    if $k = a$, area of inner loop $= a^2(\pi - \frac{3}{2}\sqrt{3})$,

             area of space between loops $= a^2(\pi + 3\sqrt{3})$.

5. The limaçon is the conchoid of a circle of radius a with respect to a point on the circumference, the fixed distance being k.

6. It is the pedal of a circle of radius k with respect to a point whose distance from the centre is $2a$.

7. It is the inverse of a conic with respect to a focus of the conic; an ellipse for $k > 2a$, a parabola for $k = a$, a hyperbola for $k < 2a$.

8. It is the epitrochoid of a point fixed at a distance a from the centre of a circle of radius $\frac{1}{2}k$ rolling on an equal fixed circle.

Roberval, between 1630 and 1640, developed a method of drawing tangents by considering a curve as being described by the resultant of two or more simultaneous movements. One of his examples is the conchoid of a circle, which he calls the 'limaçon' (i.e. snail) 'de monsieur Pascal'. This refers to Étienne Pascal, the father of Blaise Pascal. He was one of Mersenne's correspondents and the famous geometers of the day used to meet in his house.

The curve had been drawn, as described above, by Dürer, though without the name. It appears in his *Underweysung der Messung*, published in 1525.

(For Roberval's method, see the Mathematical Association Report on *The Teaching of Calculus in Schools*, p. 75.)

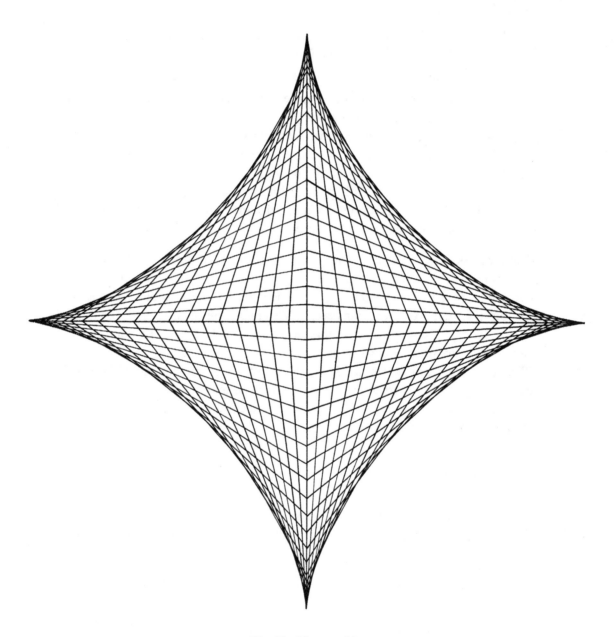

Fig. 33. The astroid

6

THE ASTROID

To Draw an Astroid (First Method)

Draw rectangular axes $X'OX$, $Y'OY$; draw, in many positions, a line QR, of fixed length $4a$, having Q on $X'OX$ and R on $Y'OY$ (Fig. 34). The envelope of the line QR is the astroid.

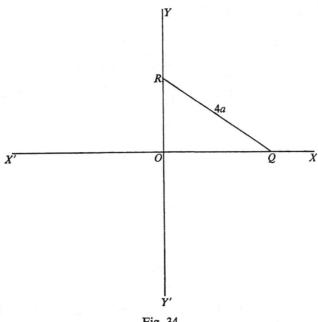

Fig. 34

Suitable Dimensions

The axes should intersect in the middle of the paper and the length $4a$ should be less than half the width of the paper. It is convenient to use one edge of a set square for drawing QR, with the length of that edge as $4a$.

To Draw an Astroid (Second Method)

Draw axes as before and draw a circle with centre O and radius $2a$, cutting OX at D and OX' at D'. Mark points on this circle at intervals of $5°$, starting from D, and number them 0, 1, 2, 3, ... (the point *D* being numbered 0), in anticlockwise order (Fig. 35). Number every third point again, starting at *D'*, in clockwise order, the intervals 0 to 1, 1 to 2, etc., being now 15° instead of 5°. Join the pairs of points having the same numbers, continuing until the whole astroid is formed.†

Suitable Dimensions

It is convenient to use a semicircular protractor for marking the intervals, the radius of the base-circle being just greater than that of the protractor. The width of

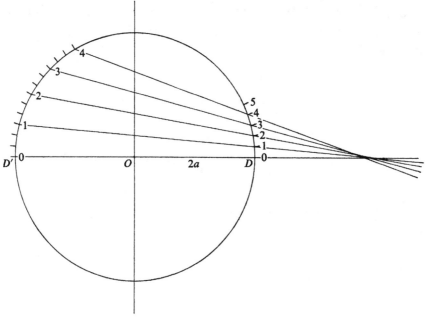

Fig. 35

the paper should be at least double the diameter of this circle. (If it is not quite wide enough the difficulty may be avoided by drawing the axes at 45° to the edges of the paper.)

Justification of the Second Method

In Fig. 36, angle $DOE = \alpha$ and angle $D'OF = 3\alpha$. Find in succession angles *FOE*, *OEF* and *EQO*, and hence prove that *RQ* is of constant length.

---

† Care is required when the numbers 'cross over'. The double numbering of the points should be continued until this has happened twice (i.e. at numbers 9 and 27). In later drawings it may be found more convenient to use a single numbering of consecutive points from 0 to 71, and to join points 0 and 36, 1 and 33, 2 and 30, etc.

(Angle $FOE = 180° - 4\alpha$. Triangle FOE is isosceles, and therefore

angle $OEF = 2\alpha$.

Angle $EQO = \alpha$, from which $EQ = EO = ER$. Therefore $QR = 4a$.)

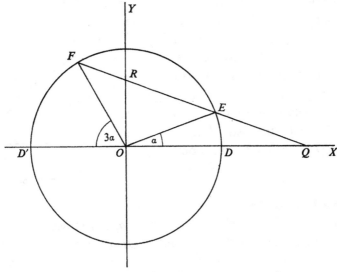

Fig. 36

Instantaneous Centre of the Line QR

Any displacement of a rigid body in a plane can be effected by means of a rotation about some point. For if, in Fig. 37, the point of the body which was at A has moved to B, and the point which was at B has moved to C, then $AB = BC$; and if O is the centre of the circle passing through A, B and C, triangles OAB and OBC are congruent and angles AOB, BOC are equal. Thus a rotation about O would bring A and B to their new positions. But if A and B are in their new positions, so is the whole body.

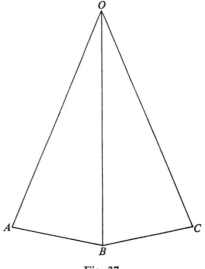

Fig. 37

In a small displacement, such as the movement of the line QR to a position $Q'R'$, the centre of rotation will be at the intersection of the perpendicular bisectors of QQ' and RR' (Fig. 38). The smaller the movement, the more nearly will this point coincide with the point of intersection of the perpendiculars drawn to the axes through Q and R. This point is called the *instantaneous centre,* because at that instant the movement of the line QR can be regarded as a rotation about that point.

In Fig. 39, I is the instantaneous centre of the line QR as Q moves along OX and R along YO. If IP is drawn at right angles to RQ, the point of the moving line which is at P must move at right angles to PI, that is, along RQ. Any other point of the moving line will move in a different direction. But the point of contact of RQ with the astroid must move along RQ. Therefore P is that point of contact.

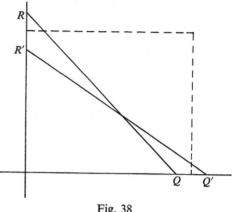

Fig. 38

The Astroid as a Hypocycloid

In the rectangle $OQIR$ (Fig. 39), $OI = QR = 4a$, so I moves on a circle, centre O, radius $4a$. Let that circle cut OX at A, and let OI cut RQ at M. Then a circle drawn

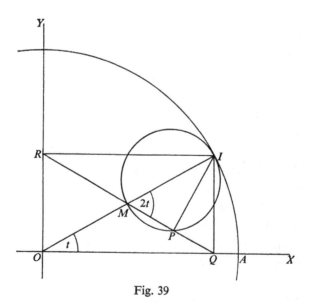

Fig. 39

on MI as diameter will pass through P, and will touch the larger circle at I. Moreover the arc PI of this circle is equal to the arc AI of the circle on which I moves (for if angle $AOI = t$, angle $PMI = 2t$ and arc PI subtends an angle $4t$ at the centre, the radius being $\frac{1}{4}$ that of the larger circle). Thus the locus of P is that of a point on the circumference of a circle rolling on the inside of a fixed circle. A curve formed in this way is called a *hypocycloid*.

Double Generation

* Let IP be produced to meet the large circle at I' (Fig. 40), and let $I'O$ produced meet QR produced at N. Then angle $OII' = 90° - 2t$ and, in the isosceles triangle OII',

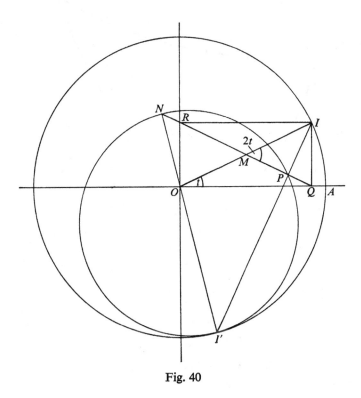

Fig. 40

angle $OI'I = 90° - 2t$. Hence angle $I'OI = 4t$, angle $I'OA = 3t$, and angle $I'NQ = 2t$. Therefore $ON = OM = 2a$, and the circle on $I'N$ as diameter will be of constant radius $3a$ and will pass through P. The arc PI' subtends an angle $4t$ at the centre of a circle of radius $3a$; and the arc AI' subtends $3t$ at the centre of a circle of radius $4a$. Therefore arc $PI' = $ arc AI', and the locus of P is that of a point on the circumference of a circle of radius $3a$, rolling on the inside of a circle of radius $4a$. Thus the astroid may be generated as a hypocycloid in two ways.

[57]

Parametric Equations for the Astroid

In Fig. 39,
$$OI = 4a,$$
$$RI = 4a\cos t,$$
$$RP = RI\cos t = 4a\cos^2 t,$$
and the distance of P from $OY = RP\cos t = 4a\cos^3 t.$

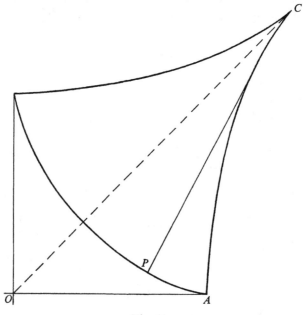

Fig. 41

Similarly,
$$QI = 4a\sin t,$$
$$QP = QI\sin t = 4a\sin^2 t,$$
and the distance of P from $OX = QP\sin t = 4a\sin^3 t.$

Therefore, if the coordinates of P are (x, y),
$$x = 4a\cos^3 t \quad \text{and} \quad y = 4a\sin^3 t.$$

Further Exercises

1. The line $I'PI$ (Fig. 40) is a normal to the astroid at P. Use the fact (proved above, p. 57) that angle $AOI' = 3 \times$ angle AOI to draw a number of normals. The envelope of these normals is the evolute of the astroid. Any point on it is the *centre of curvature* of the astroid at the point from which the corresponding normal is drawn, i.e. if PI touches the evolute at E (Fig. 42), then E is the centre, and EP is the radius, of the circle which fits most closely to the astroid in the neighbourhood of P.

[58]

* 2. If $D'OD$ is drawn (Fig. 42) bisecting angle XOY, meeting at D' and D the circle whose centre is O and whose radius is OA, with D between OX and OY, prove that angle $I'OD' = 3 \times$ angle IOD. Hence prove that the evolute of the astroid is another astroid. (*Hint:* Consider the second method of drawing the astroid.)

* 3. If C is a cusp of the evolute, imagine a string laid along the evolute from C to A, one end being fixed at C. If this string is gradually unwrapped from the

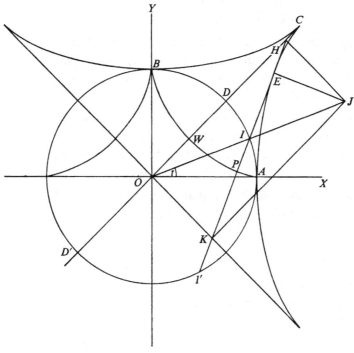

Fig. 42

evolute, as shown in Fig. 41, the point originally at A will move along the arc of the astroid.† Use this to prove that the whole length of the evolute is $48a$, and hence that the whole length of the original astroid is $24a$. (*Hint:* Consider the position when the string lies along CW.)

Area of the Astroid

* In Fig. 42, PI is a normal to the astroid, touching the evolute at E and meeting the axes of the evolute at H and K. Then $HK = 8a$ (see Ex. 2, above). $KOHJ$ is a rectangle and JE is perpendicular to HK.

† This method of relating the original curve to its evolute is explained more fully on pp. 83, 84.

Then $\qquad$ angle $JIE = 2 \times$ angle IOC

$$= 90° - 2t.$$

Hence $\qquad$ $IE = 4a\sin 2t;$

but, in Fig. 40, $\qquad$ $IP = 2a\sin 2t,$

$$\therefore \quad IE = 2 \times IP = \tfrac{2}{3}EP.$$

Now consider Fig. 43, showing PIE and another such line *pie*, these two lines meeting at F.

$$\frac{\Delta FIi}{\Delta FPp} = \frac{FI.Fi}{FP.Fp},$$

and, as p approaches P, this ratio approaches EI^2/EP^2, or $\tfrac{4}{9}$. It follows that the area bounded by the arcs AC, BC and the circular arc AIB is $\tfrac{4}{9}$ of that bounded by the three astroidal arcs AC, BC and APB.

Let the area of the original astroid be S. The evolute is a similar astroid on double scale, so its area is $4S$. Of the space between, of area $3S$, $\tfrac{4}{9}$ lies outside the circle and $\tfrac{5}{9}$ inside,

$$\therefore \quad S + \tfrac{5}{9}.3S = \pi(4a)^2 \quad \text{and hence} \quad S = 6\pi a^2.$$

The area of the astroid is thus $\tfrac{3}{8}$ that of its circumscribed circle, or $\tfrac{3}{2}$ times that of its inscribed circle.

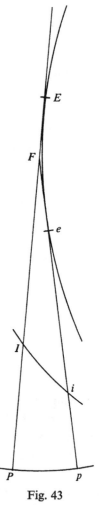

Fig. 43

The Astroid: Summary

** 1. Parametric equations are $x = a\cos^3 t, y = a\sin^3 t$. (The length $4a$ is now renamed a.)

2. The Cartesian equation is $x^{\frac{2}{3}} + y^{\frac{2}{3}} = a^{\frac{2}{3}}$.

3. $\psi = 180° - t$.

4. The pedal equation is $r^2 = a^2 - 3p^2$.

5. $A = \tfrac{3}{8}\pi a^2$.

6. $L = 6a;$ $s = \tfrac{3}{2}a\sin^2 t$, where $0 < t < 90°$.

7. $\rho = -|\tfrac{3}{2}a\sin 2t|$.

8. The evolute is another astroid, on double the scale of the original, with its axes at $45°$ to those of the original. Parametric equations for the point E (Fig. 42), using the same axes as the original and the same parameter t, are

$$x = a\cos t.(1 + 2\sin^2 t), \quad y = a\sin t.(1 + 2\cos^2 t).$$

9. The length of the tangent, measured between the axes, is constant and equal to a.

10. The astroid is the envelope of a line of fixed length a, sliding with its ends on two rectangular axes.

11. It is the hypocycloid formed by rolling a circle of radius $\frac{1}{4}a$ or $\frac{3}{4}a$ on the inside of one of radius a.

12. It is the envelope of ellipses having the same centre and orientation, the sum of their axes being constant.

The astroid seems to have acquired its present name only in 1838, in a book published in Vienna; it went, even after that time, under various other names, such as 'cubocycloid', 'paracycle', 'four-cusp-curve', and so on. The equation $x^{\frac{2}{3}} + y^{\frac{2}{3}} = a^{\frac{2}{3}}$ can, however, be found in Leibniz's correspondence as early as 1715.

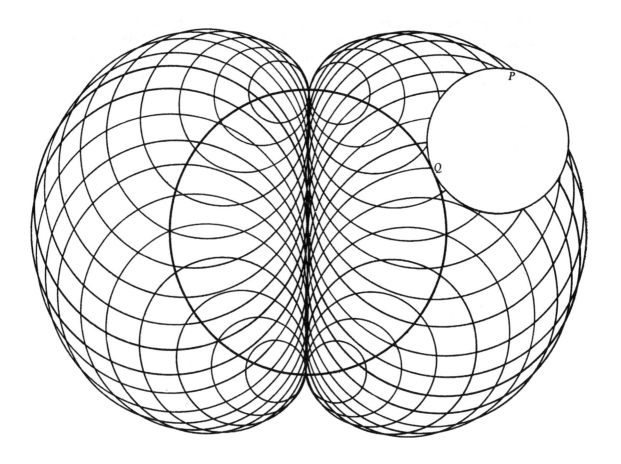

Fig. 44. The nephroid

7

THE NEPHROID

To Draw a Nephroid (First Method)

Draw rectangular axes OX, OY, and with centre O draw a circle (to be called the *base-circle*). With any point Q on the base-circle as centre, draw a circle to touch the y-axis. Repeat for many positions of Q. The envelope of these circles is the nephroid (Fig. 44).

Suitable Dimensions

With O in the centre of the paper, the radius $2a$ of the base-circle should be as follows:

Paper:	1_L	2 in.	or 5 cm.
	2_L	2·5 in.	6 cm.
	3_L	3 in.	8 cm.

The Nephroid as an Epicycloid

Let Q, q (Fig. 45) be two positions of Q, with QN, qn the perpendiculars drawn to OY. Let the circles whose centres are at Q and q intersect at P, and let Qq produced meet the axis OY at R. Then, in the similar triangles QNR, qnR,

$$\frac{QR}{qR} = \frac{QN}{qn} = \frac{QP}{qP} \text{ (constr.).}$$

Therefore, in triangle QPq, R divides the base Qq externally in the ratio of the other two sides; hence PR is the external bisector of the angle QPq. If now q approaches Q, PR will in the limit be at right angles to PQ. At the same time, P will become a point on the nephroid; triangles QNR, QPR will be congruent; and the angles QRN, QRP will be equal. Qq will become the tangent to the circle at Q, and PR will be the tangent to the nephroid.

Let the base-circle cut OY at A (Fig. 46), and let OQ produced and RP produced meet at T. Then triangles RQT, RQO are congruent, and $QT = QO$. The circle QPT, on QT as diameter, will thus be of fixed radius a, and its arc QP will be equal to the arc QA of the base-circle (angle at centre double, radius half). Therefore, if this circle rolls round the outside of the base-circle, the point P will eventually arrive at A. The nephroid, which is the locus of P, is therefore that of a point fixed on the rolling circle, namely the point which will eventually reach A.

The nephroid may, therefore, be described as the locus of a point on the circumference of a circle of radius a rolling round the outside of a fixed circle of radius $2a$.

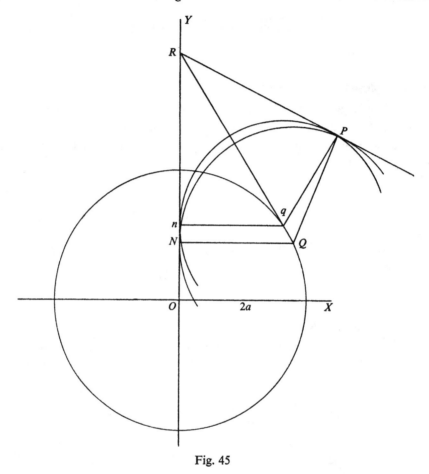

Fig. 45

It should be noted that, in Fig. 46, Q is the instantaneous centre of the rolling circle, and P moves at right angles to QP, i.e. along RT. RT is thus the tangent to the nephroid at P and PQ is the normal.

Parametric Equations for the Nephroid

* If angle XOQ is t, and the centre of the rolling circle is O' (Fig. 46), the coordinates of O' are $(3a\cos t, 3a\sin t)$. Angle $PO'T = 2t$, and the inclination of $O'P$ to OX is $3t$. The coordinates (x,y) of P are therefore given by

$$x = 3a\cos t + a\cos 3t, \quad y = 3a\sin t + a\sin 3t.$$

[64]

For the point diametrically opposite to P on the rolling circle,

$$x = 3a\cos t - a\cos 3t, \quad y = 3a\sin t - a\sin 3t.$$

The first pair of equations represent a nephroid orientated as in Fig. 44, the second pair one with its cusp-line on the axis of x.

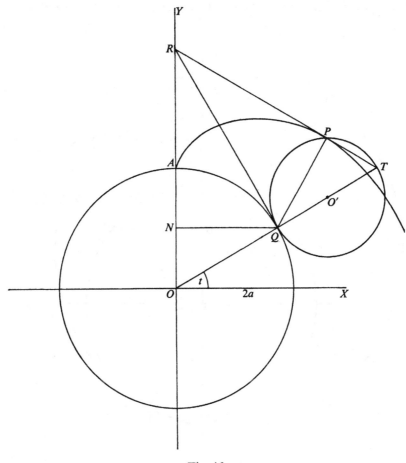

Fig. 46

Pedal Equation of the Nephroid

* In Fig. 46, angle $TQP = t$. If p is the perpendicular distance from O to PR, $p = 4a\cos t$. Applying the cosine formula to triangle OPQ,

$$r^2 = OP^2 = (2a)^2 + (2a\cos t)^2 + 2(2a)(2a\cos t)\cos t$$

$$= 4a^2(1 + 3\cos^2 t),$$

$$\therefore \quad 4r^2 - 3p^2 = 16a^2.$$

Double Generation

* The nephroid may be described as an epicycloid in another way. Let PQ be produced to meet the base-circle again at U (Fig. 47). Join UO and produce it to meet PR produced at V. Draw OM perpendicular to QU. Then triangles OMU, OMQ, TPQ, ONQ are all congruent to each other. Therefore $UP = 3UM$ and,

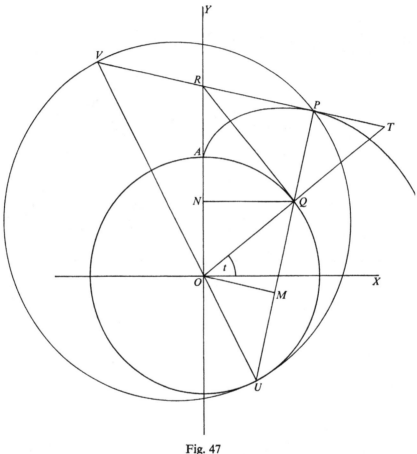

Fig. 47

since OM is parallel to VP, $UV = 3UO = 6a$. Hence the circle on UV as diameter, which passes through P, is of fixed radius $3a$. Now consider the arc UP of this circle and the arc UA of the base-circle. The angle V is $90° - t$, and therefore the arc UP subtends $180° - 2t$ at the centre of a circle of radius $3a$. But angles UOM, MOQ, QOA are each $90° - t$; so the arc UA subtends $270° - 3t$ at the centre of a circle of radius $2a$. These arcs UP and UA are therefore equal, and it follows that, if the circle on UV rolls on the outside of the base-circle, P will eventually coincide

[66]

with A. The nephroid may thus be described as the locus of a point on the circumference of a circle of radius $3a$ rolling, so as to make internal contact, on the outside of a circle of radius $2a$.

Further Exercises

1. *To draw a nephroid* (*second method*). Draw rectangular axes OX, OY, and a circle with centre O. With centre at any point R on OY, and radius RO, draw an arc to cut the circle at T (Fig. 48). Join RT. The envelope of RT, as the position of R varies, is the nephroid.

Suitable dimensions: With the paper in the 'portrait' position and the origin O in the centre, the radius of the circle should be as large as possible. It is best to

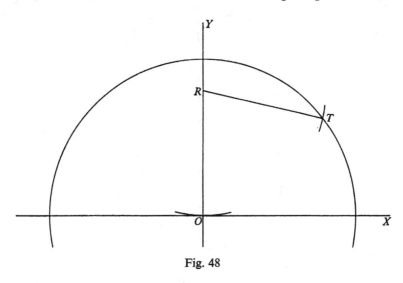

Fig. 48

begin with RO just greater than half the radius of the circle and to increase RO slowly at first, more rapidly later. It is not necessary for RO to be limited to the span of the compasses, as T can be found by the use of a ruler only.

2. *To draw a nephroid* (*third method*). Prove that, in Fig. 47,

$$\text{angle } XOV = 3 \times \text{angle } XOT.$$

(*Hint:* Use the isosceles triangle OVT.) This shows how the tangent VT may be constructed by joining points on a circle of radius $4a$, as follows:

Draw a circle, and a radius OX. Beginning at X, mark points on the circumference at intervals of $10°$ and number them 0, 1, 2, 3, ... (X itself being numbered 0). Join the points 1 and 3; 2 and 6; 3 and 9; and so on. The envelope of these joining lines is the nephroid.

Suitable dimensions: The circle should be as large as possible.

[67]

3. In Fig. 47, QU is the normal to the nephroid at P. Use the fact that

$$\text{angle } AOU = 3 \times \text{angle } AOQ$$

to draw a number of such normals. The envelope of these normals is the evolute of the nephroid.

4. *Evolute.* The similarity of the methods used in Exercises 2 and 3 above, shows that the evolute is another nephroid, on half the scale of the original one, its cusps lying on the x-axis (Fig. 49).

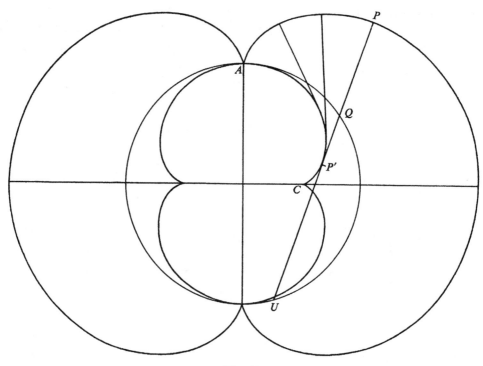

Fig. 49

5. *Maximum length and width.* The maximum length of the nephroid, measured along OX is $8a$; and the maximum width, measured parallel to OY, is $4\sqrt{2}.a$. (*Hint:* For the maximum width, QP must be parallel to OY, therefore $t = 45°$.)

6. *Arc-length.* Imagine a string laid along the evolute from the cusp C to the point A where it is crosses the y-axis. If this string is gradually unwrapped from the evolute, as shown in Fig. 49, the point starting at A will trace out the original nephroid. Use this to prove that the total arc-length of the evolute is $12a$, and that that of the original nephroid is $24a$.

7. In Fig. 47, prove that $PT = \frac{1}{4}VT$. (*Hint:* Use the similar triangles UPV, QPT.)

[68]

* 8. *Area.* The last exercise shows that the tangent to the nephroid, regarded as a chord of the circumscribing circle, is divided in the ratio $3:1$ at the point of contact. Thus if UQ (Fig. 49) touches the evolute at P', $P'Q = \frac{1}{4}UQ = \frac{1}{2}QP = \frac{1}{3}P'P$. Calling the area of the evolute S, so that that of the original nephroid is $4S$, prove that $4S = 12\pi a^2$. (*Hint:* Follow the method given for the area of the astroid, p. 59. Since $P'Q = \frac{1}{3}P'P$, the area between the evolute and the circle is $\frac{1}{3}$ that between the evolute and the original nephroid.)

9. Use a set square to plot the pedal curve of a nephroid with respect to its centre, i.e. the locus of the foot of the perpendicular from the centre to a tangent. This is a curve with two loops, facing inwards along the cusp-line and touching each other.

The Nephroid: Summary

** 1. Parametric equations for the curve as shown in Fig. 50 are

$$x = a(3\cos t - \cos 3t),$$
$$y = a(3\sin t - \sin 3t).$$

2. $\psi = 2t$.
3. The pedal equation is $4r^2 - 3p^2 = 16a^2$.
4. $A = 12\pi a^2$.
5. $L = 24a$; $s = 6a(1 - \cos t)$, where $0 < t < 180°$.
6. The intrinsic equation is $s = 6a(1 - \cos\frac{1}{2}\psi)$, where $0 < \psi < 360°$.
7. $\rho = |3a\sin t|$.
8. The evolute is a nephroid on half the linear scale, with its cusp-line at right angles to that of the original. Its parametric equations are

$$x = \tfrac{1}{2}a(3\cos t + \cos 3t), \quad y = \tfrac{1}{2}a(3\sin t + \sin 3t).$$

(*Hint:* For a point on the evolute,

$$x = a(3\cos t - \cos 3t) - \rho\sin\psi,$$
$$y = a(3\sin t - \sin 3t) + \rho\cos\psi.$$

It is sufficient to consider only values of t between $0°$ and $180°$; by symmetry the result holds for the other half of the curve. This avoids the difficulty caused by the fact that s decreases as t increases from $180°$ to $360°$.)

9. The nephroid is the epicycloid formed by a circle of radius a rolling, with external contact, on a fixed circle of radius $2a$; or by a circle of radius $3a$ rolling, with internal contact, on a fixed circle of radius $2a$.

10. It is the envelope of the diameter of a circle which rolls on the outside of an

equal circle. (*Hint:* In Fig. 46, consider a circle centre T, radius TQ, cutting TR at D. Prove that arc DQ = arc QA. Then, if the circle whose centre is T rolls on the base-circle, D will arrive at A. This shows that TD is a radius fixed relative to the rolling circle.)

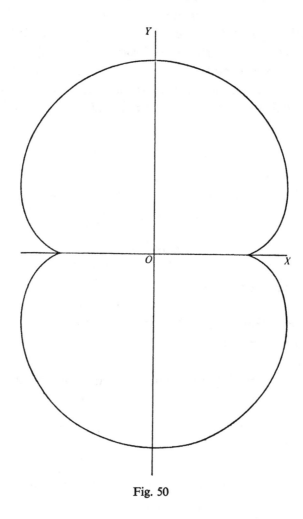

Fig. 50

11. It is the caustic of a circle for parallel rays. (*Hint:* In Fig. 46, consider the circle centre O, radius OT. If TM is the perpendicular from T to OX, OT bisects angle PTM.)

(A good approximation to part of a nephroid, formed in this way, can be seen by placing a dark-coloured cylindrical saucepan on the ground so that the rays from the sun or from a powerful electric lamp fall on it at an angle of about 60° to the horizontal.)

12. It is the caustic of a cardioid with radiant-point at the cusp. (*Hint for proof:* In Fig. 24, PA and PO' make equal angles with PQ. Therefore PO' is the reflected ray. But PO' is a diameter fixed relative to the rolling circle. Therefore the envelope is a nephroid, by no. 11, above.)

The name *nephroid* ('kidney-shaped') was used for the two-cusped epicycloid by Proctor in 1878; a year later, Freeth used the same name for a somewhat more elaborate curve (see p. 135).

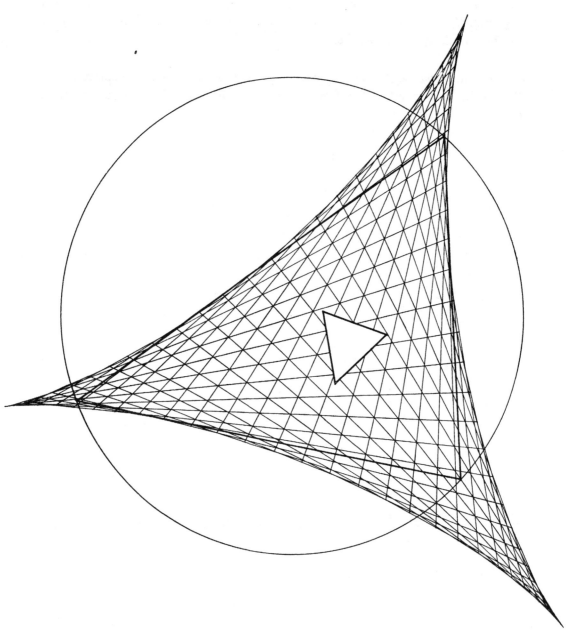

Fig. 51. Envelope of Simson's line with Morley's triangle

8

THE DELTOID

To Draw a Deltoid (First Method)

Draw a circle, centre O, and a diameter $D'OD$. Mark points on this circle at intervals of 5°, starting from D, and number them 0, 1, 2, 3, ..., in anticlockwise order, the point D itself being numbered 0. Number alternate points again, starting from D', in clockwise order, the intervals 0 to 1, 1 to 2, etc., being now 10° instead of 5°. Join the pairs of points having the same numbers, continuing until a three-cusped curve is completed. This curve is the *deltoid*.† (Fig. 51)

Suitable Dimensions

It is again convenient to use a semi-circular protractor. If its radius is 2 in. the centre should be $3\frac{1}{4}$ in. from the left-hand edge of the paper. With paper 3_P the whole curve can then be drawn, but with narrower paper one cusp will be cut off.

Envelope of Diameter of a Rolling Circle

In Fig. 52, a circle of radius a, on $D'OD$ as diameter, is drawn as before, and Q', Q are a pair of corresponding points, with angles DOQ, $D'OQ'$ equal to t and $2t$ respectively. A circle is drawn with centre O and radius $3a$, meeting $D'OD$ produced at A' and A. A variable circle is now drawn with centre Q and radius $2a$, touching the outer fixed circle at I and the inner fixed circle at J. $Q'Q$ is produced to form a diameter $P''P'$ of this circle.

Angle $P'QI = \frac{3}{2}t$. Hence the arc $P'I$, which subtends an angle $\frac{3}{2}t$ at the centre of a circle of radius $2a$, is equal to the arc AI, which subtends an angle t at the centre of a circle of radius $3a$. It follows that, if the circle on JI as diameter were rolled round the inside of the outer fixed circle, P' would reach the position A. Thus $P''P'$ is a fixed diameter of the rolling circle, and the deltoid, as drawn above, may be defined as the envelope of a diameter of a circle of radius $2a$ rolling round the inside of a fixed circle of radius $3a$.

The Deltoid as a Hypocycloid

The instantaneous centre of the rolling circle is at I. If IP is drawn at right angles to the diameter $P''P'$ (Fig. 52), P will move at right angles to PI, i.e. along the

† See footnote, p. 54.

[73]

diameter. Every other point on the diameter will be moving in a different direction; hence P is the point of contact of the diameter with its envelope the deltoid.

Consider now the circle on IQ as diameter. It will pass through P, and its arc IP (subtending an angle $3t$ at the centre of a circle of radius a) will be equal to the arc IA (subtending an angle t at the centre of a circle of radius $3a$). Hence, if the circle on IQ (shown in Fig. 53) rolls on the inside of the outer fixed circle, P will arrive at A. The deltoid, the locus of P, is therefore a hypocycloid.

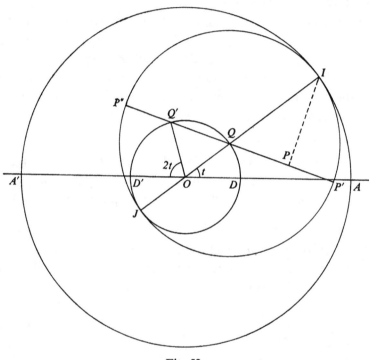

Fig. 52

* Now produce IP to meet the outer circle at I' (Fig. 53). Then $I'OQ'$ is a straight line (for angle $I = 90° - \frac{3}{2}t$; therefore angle $I'OI = 3t$, and angle $I'OA = 2t$). The circle on $I'Q'$ as diameter will therefore have a fixed radius $2a$ and will pass through P. Moreover, its arc $I'P$ (angle at centre $3t$, radius $2a$) is equal to the arc $I'A$ of the fixed circle (angle at centre $2t$, radius $3a$). Hence the locus of P is that of a point on the circumference of a circle of radius $2a$ rolling on the inside of a fixed circle of radius $3a$.

Thus the deltoid may be generated as a hypocycloid in two ways, the rolling circle having a radius equal to either $\frac{1}{3}$ or $\frac{2}{3}$ that of the fixed circle.

[74]

Motion of Diameter of the Rolling Circle

*   It was seen that, in Fig. 52, P' was a fixed point on the circumference of the rolling circle, the radius of that circle being $\frac{2}{3}$ that of the fixed circle. Its locus is therefore a deltoid and, since it reaches the fixed circle at A, the cusps of this deltoid must coincide with those of the locus of P, and the envelope of the diameter $P''P'$. The same applies to the other end of the diameter, P'', the cusps of its locus being in the same positions, since the semicircular arc $P'P''$ is of length one-third that of the

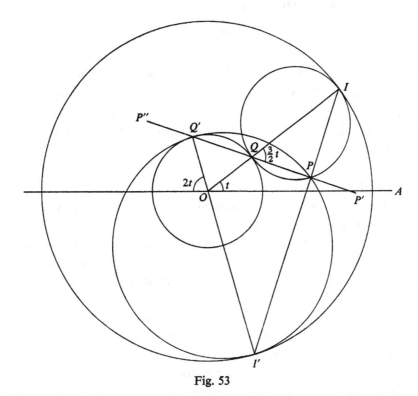

Fig. 53

circumference of the fixed circle. (After the circle has rolled once round, P' and P'' will have changed places.)

Thus P, P' and P'' all have the same locus, which is also the envelope of the diameter $P''P'$. This diameter in fact moves with its two ends on the deltoid which it touches.

To Draw a Deltoid (Second Method)

Proceed as in the first method; but, on joining any pair of points Q, Q', produce the line both ways and mark points on it at a distance $2a$ from Q. The deltoid will be formed partly as an envelope, partly as a locus.

[75]

Parametric Equations

* In Fig. 52, QP' makes an angle $\frac{1}{2}t$ with OA. If OA is taken as axis of x and a line through O at right angles to it as axis of y, the coordinates (x,y) of P' are given by

$$x = OQ\cos t + QP'\cos\tfrac{1}{2}t = a\cos t + 2a\cos\tfrac{1}{2}t;$$
$$y = OQ\sin t - QP'\sin\tfrac{1}{2}t = a\sin t - 2a\sin\tfrac{1}{2}t.$$

**The coordinates of P are likewise given by

$$x = OQ\cos t + QP\cos\tfrac{1}{2}t$$
$$= a\cos t + 2a\cos\tfrac{3}{2}t\cos\tfrac{1}{2}t = a\cos 2t + 2a\cos t;$$
$$y = OQ\sin t - QP\sin\tfrac{1}{2}t$$
$$= a\sin t - 2a\cos\tfrac{3}{2}t\sin\tfrac{1}{2}t = -a\sin 2t + 2a\sin t.$$

It will be noticed that the second pair of equations can be obtained from the first pair by changing t into $-2t$. This means that, if the rolling circle, starting from A, were to go twice as far in the opposite direction, P' would reach the position occupied by P.

Geometrical Properties

1. The tangent to the deltoid, measured between the two points in which it cuts the curve again, is of constant length $4a$.

2. The tangents to the deltoid at P' and P'' (Fig. 54) meet at right angles at a point on the inner fixed circle. (*Hint:* Since I is the instantaneous centre of the rolling circle, these tangents are at right angles respectively to $P'I$ and $P''I$, and meet at the point called J in Fig. 52.)

3. The normals at P, P' and P'' meet on the outer fixed circle.

*   4. Suppose that, in Fig. 53, a line $E'OE$ is drawn so that angle $AOE = 60°$ (measured anticlockwise). Prove that angle $I'OE' = 2 \times$ angle IOE. Hence prove that the evolute of a deltoid is another deltoid.

*   5. Prove that the whole length of the evolute is $48a$, and that of the original deltoid $16a$. (*Hint:* Use the method suggested for the astroid, p. 59.)

Further Drawing Exercises

*   1. Draw a circle and inscribe in it a triangle, which should be neither right-angled nor isosceles. Take any point P on the circle and use a set square to mark the feet of the perpendiculars from it to the sides of the triangle. (The perpendiculars should not be drawn.) Verify that the three points so found lie on a straight line. Draw this line (known as *Simson's Line*) for a large number of positions of P. Its envelope is a deltoid (Fig. 51).

Suitable Dimensions

The diameter of the circle should be less than two-thirds the width of the paper. The triangle may be acute- or obtuse-angled, but, if an obtuse-angled triangle is used, it is the triangle rather than the circle that should be in the middle. It is advisable to draw the sides of the triangle in ink. They will need to be produced,

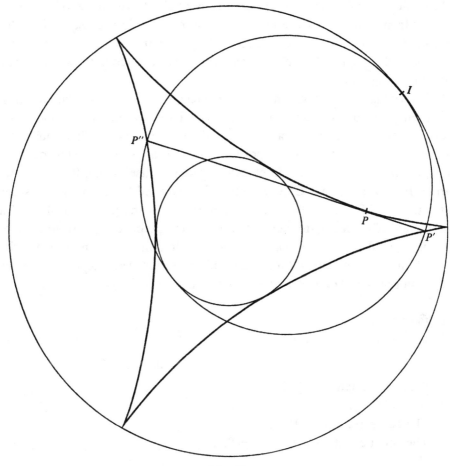

Fig. 54

and this may be done in pencil. It is best to mark five or six positions of P and draw the corresponding Simson Lines before going on to another batch.

2. Find the nine-points centre of the same triangle (i.e. the mid-point of the line joining the centre of the circle to the point where the altitudes of the triangle meet). Verify that it is the centre of the deltoid, that the lines joining it to the cusps make angles of 120° with one another, and that the nine-points circle of the triangle (radius

half that of the circumcircle) is the inscribed circle of the deltoid. Draw also the circumcircle of the deltoid, with radius $\frac{3}{2}$ times that of the circumcircle of the triangle.

(For a proof of these properties, see *The Mathematical Gazette*, vol. xxxvii, p. 124. They are illustrated in a film by T. Fletcher, of the Sir John Cass College, London, E.C. 3.)

3. In the same figure, draw the trisectors of the three angles of the triangle. Of those from two vertices B and C mark the intersection nearest to BC. Do the same for the other two pairs of vertices. The three points marked in this way form an equilateral triangle (*Morley's triangle*) which is orientated in the same way as the deltoid (see Fig. 51).

* 4. If a triangle rotates within its circumcircle, the envelope of the Simson Line of a fixed point on that circle is a cardioid. To draw this, mark points at 10° intervals round a circle. Choose one of these points as the fixed point P and three more of them to form one position of the triangle ABC. Starting from A, label the points 0, 1, 2, 3, ..., in clockwise order, A itself being 0. Do the same, starting from B and from C. The three points labelled '1' represent 'position 1' of the triangle. Place a ruler as if for joining two of the points labelled '1' and, with the aid of a set square, mark the foot of the perpendicular from P. Do the same using a different pair of points labelled '1'. Join the two points so marked: this is 'position 1' of the Simson Line and should be so labelled. Repeat for position 2 and so on (see *The Mathematical Gazette*, vol. xlv, p. 220).

The Deltoid: Summary

** 1. Parametric equations are

$$x = 2a\cos t + a\cos 2t, \quad y = 2a\sin t - a\sin 2t.$$

(These are coordinates of P in Fig. 53.)

2. $\psi = 180° - \frac{1}{2}t$.
3. The cusps are at $t = 0°, 120°, 240°$.
4. The pedal equation is $r^2 + 8p^2 = 9a^2$.
5. $A = 2\pi a^2$.
6. $L = 16a$; $s = \frac{8}{3}a(1 - \cos\frac{3}{2}t)$, where $0 < t < 120°$.
7. $\rho = -|8a\sin\frac{3}{2}t|$.
8. The evolute is a deltoid on three times the linear scale of the original, with a 60° change in orientation. Its parametric equations (referred to the same axes) are

$$x = 6a\cos t - 3a\cos 2t, \quad y = 6a\sin t + 3a\sin 2t.$$

9. The deltoid is a hypocycloid formed by rolling a circle of radius a or $2a$ on the inside of a fixed circle of radius $3a$.

10. It is the envelope of a diameter of a circle of radius $2a$ rolling on the inside of a fixed circle of radius $3a$.

11. The envelope of the Simson's Lines of a triangle is a deltoid whose inscribed circle is the nine-points circle of the triangle.

12. The tangent to the deltoid, terminated at the points where it meets the curve again, is of constant length, and its mid-point lies on the inscribed circle.

13. The tangents at those points are at right angles and meet on the inscribed circle; the normals at the same points and at the point of contact of the original tangent are concurrent, meeting on the circumscribed circle.

That the envelope of the Simson's Lines of a triangle was a three-cusped symmetrical curve was discovered by J. Steiner in 1856; the curve was soon recognized as a hypocycloid and it is indeed often referred to as *Steiner's hypocycloid*. Our name, *deltoid* ('shaped like a Greek letter Δ'), is not used everywhere.

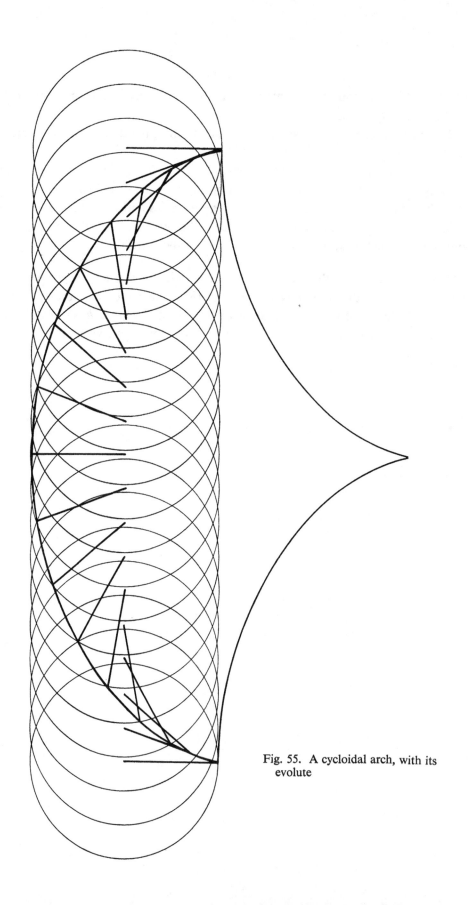

Fig. 55. A cycloidal arch, with its evolute

9

THE CYCLOID

To Draw a Cycloid (First Method)

Draw two lines 'horizontally' (i.e. across the page) at a distance a apart. Draw a series of 'vertical' lines crossing the 'horizontal' lines at intervals of $2\pi a/18$ ($= 0.35a$, approx.). Label the points on the upper 'horizontal' line 0, 1, 2, 3, ..., 18, and with these points as centres draw circles of radius a. These represent positions of a circle rolling along the lower 'horizontal' line, the intervals corresponding to turns of 20°. From the points 0, 1, 2, 3, ..., draw radii at 0°, 20°, 40°, 60°, ..., to the 'vertical' lines, the angles being measured away from the side towards which the circle is rolling (Fig. 55). These radii represent successive positions of a radius fixed on the rolling circle. Their extremities lie on the cycloid.

Suitable Dimensions

		Radius a		Intervals $0.35a$	
Paper:	1_L	1 in.	or 3 cm.	0·35 in. or 1·05 cm.	
	2_L	1·43 in.	4 cm.	0·5 in.	1·4 cm.
	3_L	2 in.	5 cm.	0·7 in.	1·75 cm.

The lower 'horizontal' line should run across the middle of the paper.

(This method of drawing the cycloid shows clearly the nature of the curve and is recommended for a first drawing. In practice the next method, which is essentially the same, will be found to be quicker.)

To Draw a Cycloid (Second Method)

Draw a 'horizontal' line and mark on it nineteen points at intervals of $0.35a$. (These will be referred to as 'points 0, 1, 2, 3, ...', counting from left to right.) With point 0 as centre and radius a draw a semicircle, to the left, and mark points on it at 20° intervals, starting from the bottom. Through these ten points draw 'horizontal' lines (to be referred to as 'lines 0, 1, 2, 3, ..., 9', the lowest one being 'line 0'). With radius a and points 0, 1, 2, 3, ..., 18 as centres, draw arcs cutting or touching lines 0, 1, 2, 3, ..., 8, 9, 8, ..., 0, respectively. (The arcs drawn with centres 0, 9 and 18 will touch the corresponding 'horizontal' line. In other cases the intersection on the side away from the centre of the diagram must be chosen.)

Parametric Equations of the Cycloid

Let an arch of the cycloid be ACB, with C the middle point (Fig. 56). Let P be any point on the curve, with OP the corresponding radius of the rolling circle. Let PM be the perpendicular drawn from P to the 'vertical' radius OI. Then, if $OP = a$ and angle MOP is t (radians), $AI = $ arc $PI = at$. Using rectangular axes ABX, AY, the coordinates (x, y) of P are given by

$$x = at - a\sin t = a(t - \sin t),$$

$$y = a - a\cos t = a(1 - \cos t).$$

These are parametric equations for the cycloid.

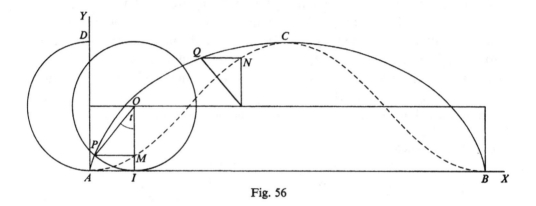

Fig. 56

Roberval's Curve

Draw successive positions of PM. The locus of M (Roberval's curve) divides each of the rectangles AC and CB symmetrically; for, if PM and QN correspond to t and $\pi - t$ respectively, it is seen that N is as far below and to the left of C as M is above and to the right of A. (These distances are, respectively, $a - a\cos t$ and at.)

The length of the base-line AB is equal to that of the circumference of the rolling circle, $2\pi a$. Therefore the area between Roberval's curve and the base-line is equal to half the rectangle $2\pi a \cdot 2a$, i.e. to $2\pi a^2$. Roberval found the area between his curve and the cycloid by observing that the width PM is everywhere equal to the width, at the same height, of the semicircle on AD as diameter (see Fig. 56). He concluded that the area between the two arcs AC was equal to that of the semi-circle, and thus found that the area of the cycloidal arch was $2\pi a^2 + \pi a^2$, i.e. $3\pi a^2$. (See pp. 84, 150, for other ways of finding this area.)

[82]

Tangent, Normal and Evolute

The instantaneous centre of the rolling circle is at I; therefore P is moving at right angles to PI, i.e. towards J, the opposite end of the diameter through I (Fig. 57). Thus PJ is the tangent to the curve at P and PI is the normal. If PI is produced to P' so that $PI = IP'$, the distances of P' to the right of and below A are the same as those of Q to the left of and below C, Q being defined as before. (These distances are, respectively, $at + a\sin t$ and $a - a\cos t$.) It follows that the locus of P' is an equal cycloid, the centre of one arch being at A.

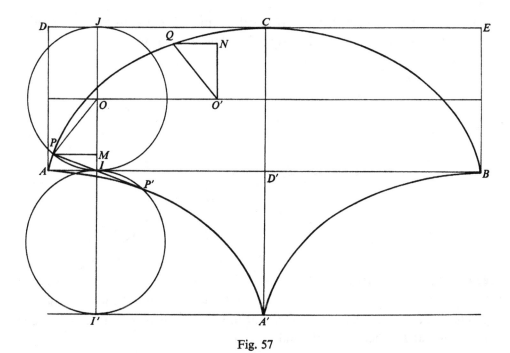

Fig. 57

This cycloid would be the locus of a point on the circumference of a circle of radius a rolling along the line $I'A'$ shown in Fig. 57, I' being the instantaneous centre when the tracing-point is at P'. The direction of motion of P' is at right angles to $P'I'$, i.e. along IP'. Thus PIP' is a tangent to the new cycloid as well as a normal to the original one; and the new cycloid is the evolute of the other.

Arc-Length

Let PP' and pp' be two normals to the cycloid at P and p (Fig. 58); let their point of intersection be R and let their points of contact with the evolute be P' and p' respectively. The nearer p is to P, the nearer will angles pPR and PpR be to right

6-2

angles, and the more nearly will RP and Rp be equal. Then $pp' - PP'$ is very nearly equal to $P'R + Rp'$, which in turn is very nearly the length of the arc $P'p'$ of the evolute. Thus if a string were laid along the evolute from p' to P' and thence along the straight line $P'P$, it could be 'unwrapped' into the position $p'p$.†

Suppose that a string is laid along the arc $A'A$ (Fig. 57) of the evolute, the end A' being fixed, and is 'unwrapped' from A. The end of the string at A will move along the original cycloid to C. It follows that the length of the half-arch $A'A$ of the evolute is equal to $A'C$, i.e. to $4a$. The length of a complete arch of either cycloid is thus $8a$.

By the same argument, the arc AP' of the evolute is equal to PP', or twice IP'. It follows that the arc PC of the original cycloid is equal to twice PJ.

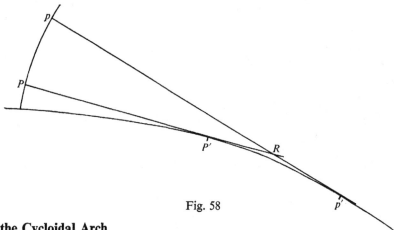

Fig. 58

Area of the Cycloidal Arch

* In Fig. 59, P and p are points on the cycloid, with P' and p' the corresponding points of the evolute, the two normals cutting the base-line at I and i, and meeting each other at R. Then $PI = IP'$ and $pi = ip'$.

$$\frac{\triangle RIi}{\triangle RPp} = \frac{RI.Ri}{RP.Rp} \text{ (by use of the formula } \triangle = \tfrac{1}{2}ab\sin C),$$

and this approaches the limiting value $\frac{1}{4}$ as p approaches P. It follows that the area $AA'B$ (Fig. 57), between the evolute and the original base-line AB, is $\frac{1}{4}$ the area $ACBA'$, between the evolute and the original cycloid. But the two parts $A'D'A$ and $A'D'B$ of the former area are congruent respectively to BEC and ADC. Therefore, these last two areas are together $\frac{1}{4}$ of the rectangle $ABED$, and the area of the cycloidal arch is $\frac{3}{4}$ of that rectangle, i.e. $\frac{3}{4} \times 2\pi a . 2a$, or $3\pi a^2$.

† This is an intuitive argument: by using the methods of the Differential Calculus it can be proved rigorously that the arc of the evolute between any two points is equal to the difference of the radii of curvature at the corresponding points of the original curve.

Motion of the Tracing-Point

** If the rolling circle turns at a uniform speed ω, the velocity of the tracing-point is $\omega.IP$, since I is the instantaneous centre (Fig. 57). This is equal to $\frac{1}{2}\omega.PP'$, or $\frac{1}{2}\omega \times$ arc AP'. The acceleration of P along the arc is therefore equal to $\frac{1}{2}\omega \times$ the velocity of P' along the arc of the evolute. But

$$\text{the velocity of } P' = \omega.P'I' = \omega.PJ = \tfrac{1}{2}\omega \times \text{arc } PC.$$

Thus the acceleration of P along the arc $= \frac{1}{4}\omega^2 \times$ arc PC. The motion of P along the arc is therefore simple harmonic.†

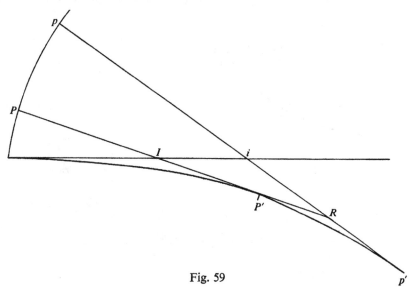

Fig. 59

Tautochrone Property of the Cycloid

** Fig. 60 is the same as Fig. 57, lettered the same way, but turned through two right angles. The angle between the tangent at P and the horizontal is called ψ. Angle PIJ is also equal to ψ. If P is a particle sliding under gravity along a smooth cycloidal arc ACB, its acceleration along the arc will be $g\sin\psi$, i.e.

$$g.\frac{PJ}{IJ} \quad \text{or} \quad \frac{g}{2a}.PJ \quad \text{or} \quad \frac{g}{4a} \times \text{arc } CP.$$

Thus the motion of P along the arc will be simple harmonic, with period $2\pi\sqrt{(4a/g)}$, the period being thus independent of the starting-point. The word *tautochrone* is used for a curve such that the time of descent from any point to the lowest point is always the same.

† Mr P. Gant has pointed out that, since the motion of O is uniform, the resultant acceleration of P is $a\omega^2$ along PO; constant in magnitude, though varying in direction.

[85]

Isochrone Property

** If, instead of sliding on a smooth curve, P is attached to A' by a string constrained to move between the cycloidal arcs $A'A$ and $A'B$, the same motion will result. This is the *cycloidal pendulum*. It has the *isochronous* property, that the time of swing is independent of the amplitude.

Further Drawing Exercises

1. *Envelope method.* This is a quicker way of drawing a cycloid than those described above. Draw a line across the paper and mark nineteen points on it at

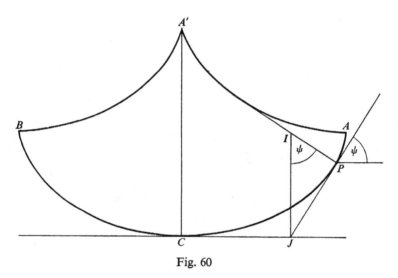

Fig. 60

intervals $0.35a$, where a is a convenient constant. With centres at the second, third, fourth, ..., points draw arcs of radii $2a\sin 10°$, $2a\sin 20°$, $2a\sin 30°$, ... Draw a freehand curve touching all these arcs.

2. Your original drawing of the cycloid shows a radius of the rolling circle in nineteen different positions. Produce these nineteen radii till each intersects its neighbours and draw a freehand curve touching all of them.

* Consider a circle on OI as diameter (Fig. 61), cutting OP at X. Prove that the arc IX is equal to IA, and that the locus of X is a cycloid to which XO is a tangent. But, as XO is part of a fixed diameter of the original rolling circle, this proves that the envelope of a diameter of a rolling circle is a cycloid.

3. Draw an approximate cycloidal arch with the aid of circular arcs, as shown in Fig. 62. (A' is the centre of curvature, with radius $4a$, for the point C of the cycloid; H and K are centres of curvature, with radius $2a$, for the points P and Q, where HP

and KQ are inclined at 30° to the base-line.) The dotted curves indicate the parts of the arch which may be drawn freehand. Alternatively, the larger joining arcs may be drawn with compasses, using radius 3·35a and centres at distances 0·25a on either side of $A'L$, at a height 0·6a above A'. Similarly the smaller joining arcs may be drawn with radius 0·8a and centres on AB.

The Cycloid: Summary

** 1. Parametric equations are $x = a(t - \sin t)$, $y = a(1 - \cos t)$.
   2. $\psi = 90° - \tfrac{1}{2}t$.

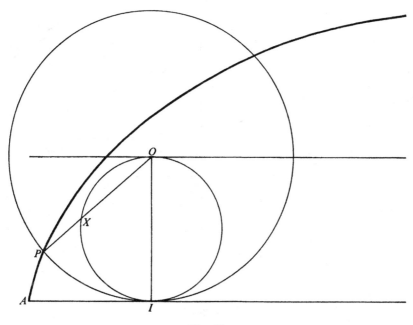

Fig. 61

3. Area between one arch and the base $= 3\pi a^2$.

4. $s = 4a(1 - \cos \tfrac{1}{2}t)$, where $0 < t < 360°$.

5. The intrinsic equation, for one arch, is $s = 4a(1 - \sin \psi)$; or, with origin at the centre of the arch, $s = 4a \sin \psi$.

6. Length of base-line from cusp to cusp $= 2\pi a$.

7. $\rho = -|4a \sin \tfrac{1}{2}t|$.

8. The evolute is an equal cycloid.

9. The cycloid is the locus of a point on the circumference of a circle rolling along a fixed straight line.

[87]

10. It is the envelope of a diameter of a circle rolling along a fixed straight line.

11. If a particle moves under gravity along a smooth curve in the form of a cycloid, placed in a vertical plane with its cusps pointing vertically upwards, the time of descent to the lowest point is independent of the starting position (the *Tautochrone Property*).

12. The cycloid is the curve down which a smooth particle will most quickly move under gravity from one given point to another the (*Brachystochrone Property*).

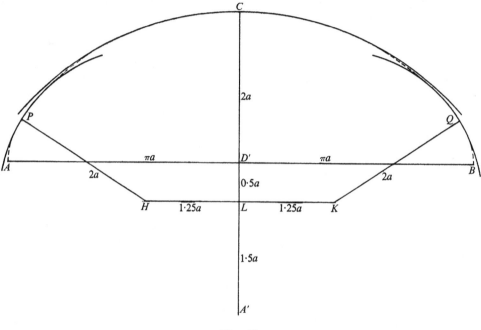

Fig. 62

The early history of the cycloid is mixed with legend, starting with Galileo's attempt to find the area. This was perhaps by weighing shaped pieces of metal, in the expectation that the ratio of cycloid to generating circle would be as π to 1. The exact area was found by Galileo's pupil Torricelli, and also by Fermat, Roberval and Descartes. Roberval and Christopher Wren succeeded in finding the length of the arc, and in 1658 Pascal offered a prize for the solution of a number of problems connected with 'la Roulette' (as it was called by the French: Galileo had probably given it the name of 'cycloid'). Only two serious competitors seem to have sent in their answers, Lalouère and Wallis. Pascal did not award the prize and his *Histoire de la Roulette* led to some sorry examples of almost nationalistic polemics. The isochrone and tautochrone properties were discovered by Huygens, and the brachy-stochrone property by James Bernoulli. The first pendulum clock, invented by Huygens, contained a device for making the pendulum isochronous by causing it to describe a cycloidal arc, using the evolute of the curve as a guide (see p. 86); but this refinement is vitiated by mechanical difficulties in construction. Cycloidal teeth for gear wheels had already been proposed by Desargues in the earlier part of the seventeenth century.

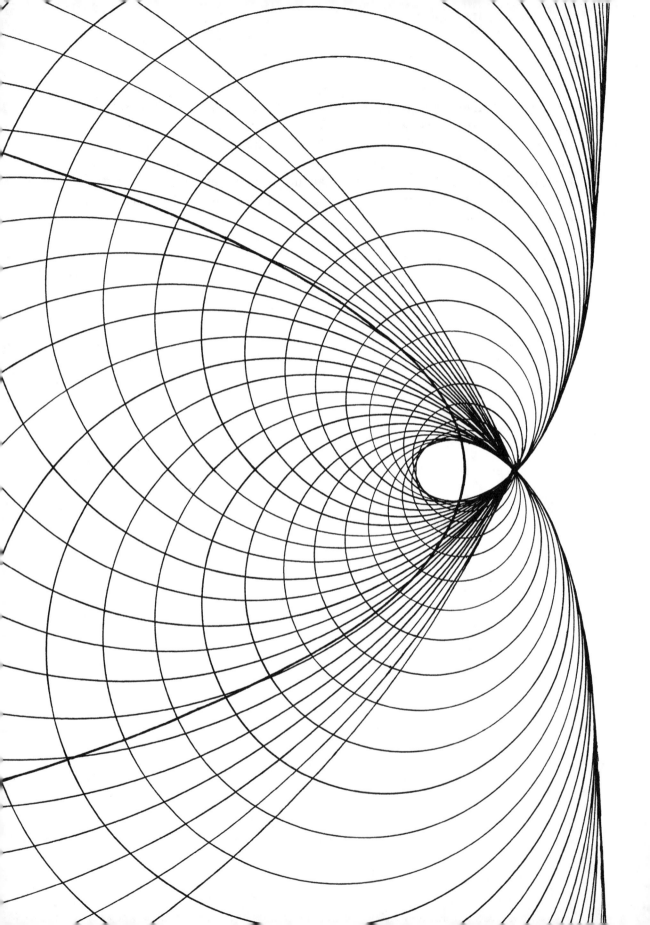

10

THE RIGHT STROPHOID

To Draw a Strophoid (First Method)

Let *PQR* be a 60° set square, with the right angle at *P* and the 60° angle at *Q*. Let *O* be a fixed point whose perpendicular distance *OA* from a fixed straight line is equal to the shortest side *PQ* of the set square. If the set square is placed with *Q* on the fixed line and *PR* passing through *O* (Figs. 64, 65), the locus of *P* is a strophoid.

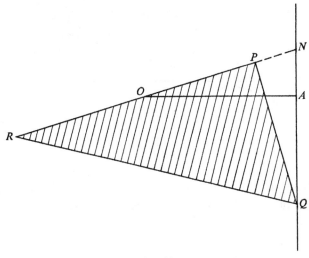

Fig. 64

Suitable Dimensions

With the paper in the 'portrait' position, the point *A* should be in the centre, with the fixed line 'vertical'. The width of the paper should be more than twice the shortest side of the set square.

To Draw a Strophoid (Second Method)

Let *OP* meet *QA* at *N* (Figs. 64, 65). Triangles *OAQ*, *QPO* are congruent and so are triangles *OAN*, *QPN*. Therefore *NP* = *NA*. This leads to the following (more accurate) method of drawing the curve:

[91]

Fig. 63. The parabola and the right strophoid

Let O be a fixed point and OA the perpendicular drawn from O to a fixed line. Draw any line through O, cutting the fixed line at N, and mark on it two points P and P' such that $NP = NP' = NA$. The locus of these points, as ON varies in direction, is the strophoid. (The word *strophoid* is used more generally for any

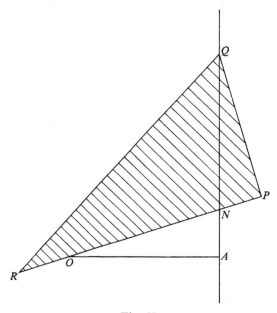

Fig. 65

curve that can be drawn in this manner. See p. 135. The particular strophoid described here is properly known as the *right strophoid*.)

Asymptote

It will be seen that, as N moves further away from A, one of the points P, P' moves nearer to O and the other approaches a straight line, parallel to the fixed line, whose distance from O is equal to twice OA. This line is an asymptote to the curve.

Polar Equation of the Right Strophoid

If OA (of length a) is taken as initial line, and angle AOP is θ, then the length r of OP is equal to $ON \pm NA$. The polar equation is, therefore,

$$r = a(\sec\theta \pm \tan\theta).$$

[92]

The Strophoid as the Pedal of a Parabola

In Figs. 64, 65, it was seen that triangles OPQ, QAO were congruent. Therefore the points O and Q, and also the points A and P, are symmetrically placed with respect to an axis passing through N. Let NT be this axis (Fig. 66), and let it cut OQ at U, AP at V. If B is the mid-point of OA, UB will be parallel to the fixed line AQ, and will itself be a fixed line. Since O is a fixed point and angle OUT is a right angle, the envelope of NT will be a parabola whose focus is O and whose directrix is the fixed line AQ (see p. 3). The point V is the foot of the perpendicular from A to

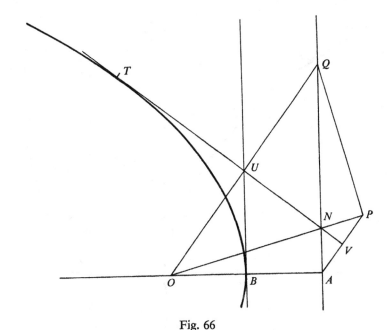

Fig. 66

the tangent NT to this parabola. The locus of V is thus the pedal of A with respect to the parabola, and the locus of P, the strophoid, is a similar curve on double scale.

Motion of the Set Square

* If T is the point of contact of NT with the parabola, TQ will be perpendicular to AQ; for $TQ = TO$ (by symmetry), and TO is equal to the perpendicular distance from T to AQ (by the focus and directrix property of the parabola).

Now consider the motion of the set square in Figs. 64, 65. Q is moving along QN and the point of the set square which is momentarily at O is moving in the direction *NO*. But angles *TQN* and *TON* are right angles; therefore the instantaneous centre is at *T*. Hence *P* is moving at right angles to *PT*, and *PT* is a normal to the strophoid at *P*.

To Draw a Strophoid (Envelope Method)

Draw a parabola with focus *O* and vertex *B*, its axis *OB* meeting the directrix at *A* (as in Fig. 66). With any point *T* on the parabola as centre, and radius *TA*, draw a circle. Repeat for numerous positions of *T*. The envelope of these circles will be a right strophoid (Fig. 63).

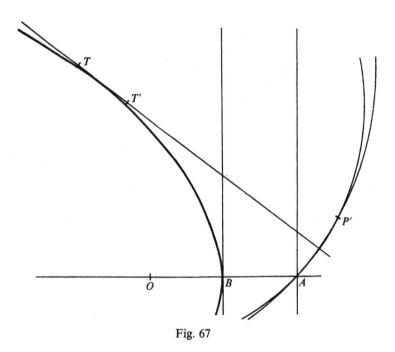

Fig. 67

Suitable dimensions: With the paper in the 'portrait' position, the point *O* should be in the centre, with *OB* 'horizontal'. The distance *OB* should be about an eighth of the width of the paper.

*   *Proof:* If *T* and *T'* are points on the parabola (Fig. 67), and the corresponding circles intersect again at *P'*, then triangles *TT'A*, *TT'P'* are congruent. Hence *P'* is the image of *A* in the chord *TT'*. As *T'* approaches *T*, the chord becomes the tangent at *T* to the parabola, and *P'* becomes the point *P* of Fig. 66.

Moreover, as *T* is the instantaneous centre of the moving set square when its right-angled corner is at *P*, the tangent to the strophoid is at right angles to *TP*, i.e. in the same direction as the tangent at *P* to the circle whose centre is *T*.

[94]

Further Drawing Exercises

1. *The oblique strophoid.* Let O be a fixed point and let A be a point on a fixed line, other than the foot of the perpendicular from O. Draw any line through O, cutting the fixed line at N, and mark on it two points P and P' such that

$$NP = NP' = NA.$$

The locus of these points, as the line through O varies, is an oblique strophoid.

2. *The right strophoid as a cissoid.* Let OX, OY be perpendicular lines through the centre O of a circle, OX cutting the circle at A (Fig. 68). Draw any line

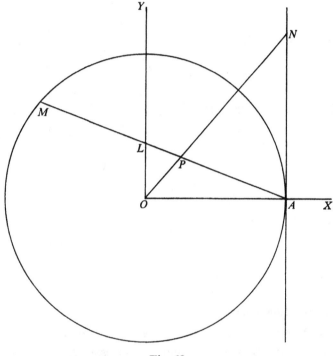

Fig. 68

through A, meeting OY at L and meeting the circle again at M. Mark a point P on this line such that $AP = LM$, the direction from A to P being the same as that from L to M. Repeat for numerous positions of the line through A. The locus of P will be a right strophoid.

A curve drawn in this manner is called a *Cissoid.* See chapter 15, p. 131.

Suitable dimensions: With the paper in the 'portrait' position, OA should be in the middle, and should be 'horizontal'. Its length should be rather less than one-third of the width of the paper.

Proof: If the tangent to the circle at A meets OP, produced if necessary, at N, triangle NAP is similar to triangle OLP. But $OL = OP$; therefore $NP = NA$.

(The oblique strophoid also may be drawn in this manner, the lines OX and OY being no longer at right angles.)

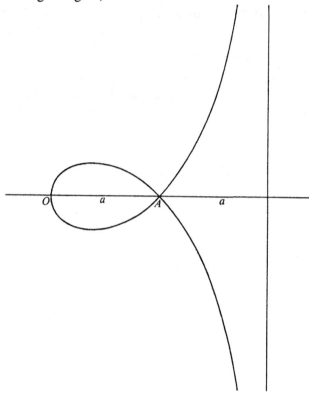

Fig. 69

The Right Strophoid: Summary

** The vertex is O and the node is A (Fig. 69). OA is used as initial line for polar coordinates and as axis of x for rectangular Cartesian coordinates.

1. The polar equations are:

(with pole at O) $r = a(\sec\theta \pm \tan\theta)$;

(with pole at A) $r = a(\sec\theta - 2\cos\theta)$.

2. The Cartesian equations are:

(with origin at O) $y^2(2a-x) = x(x-a)^2$;

(with origin at A) $y^2(a-x) = x^2(a+x)$.

3. With pole at O, $\tan\phi = \pm\cos\theta$. A line through O makes equal angles with the curve at its other two points of intersection.

[96]

4. With O as centre of inversion, and radius of inversion a, the curve inverts into itself; with A as centre, and the same radius of inversion, the inverse is a rectangular hyperbola whose vertices are at O and A.

5. Area of loop $= \frac{1}{2}a^2(4-\pi)$; area between curve and asymptote $= \frac{1}{2}a^2(4+\pi)$.

6. Parametric equations, with A as origin, are:

$$x = a \cdot \frac{t^2-1}{t^2+1}, \qquad y = at \cdot \frac{t^2-1}{t^2+1}.$$

7. The Right Strophoid is the strophoid of a straight line with respect to a pole O, with the foot of the perpendicular from O to the line as fixed point. (See p. 135 for definition of 'strophoid'.)

8. It is the pedal of a parabola with respect to the point of intersection of the axis and the directrix.

9. It is the inverse of a rectangular hyperbola with respect to a vertex.

10. It is the cissoid of a circle and one of its diameters with respect to a point on the circumference equidistant from the ends of the diameter.

The strophoid was first described in the correspondence of Torricelli, about 1645. It was first found (probably by Roberval) as the locus of the foci of a (variable) conic section when the plane cutting the cone turns about the tangent at the vertex of the conic. The modern name (which possibly means 'shaped like a *strophos*, the belt with a twist to carry a sword') comes from a French writer, Montucci, 1846.

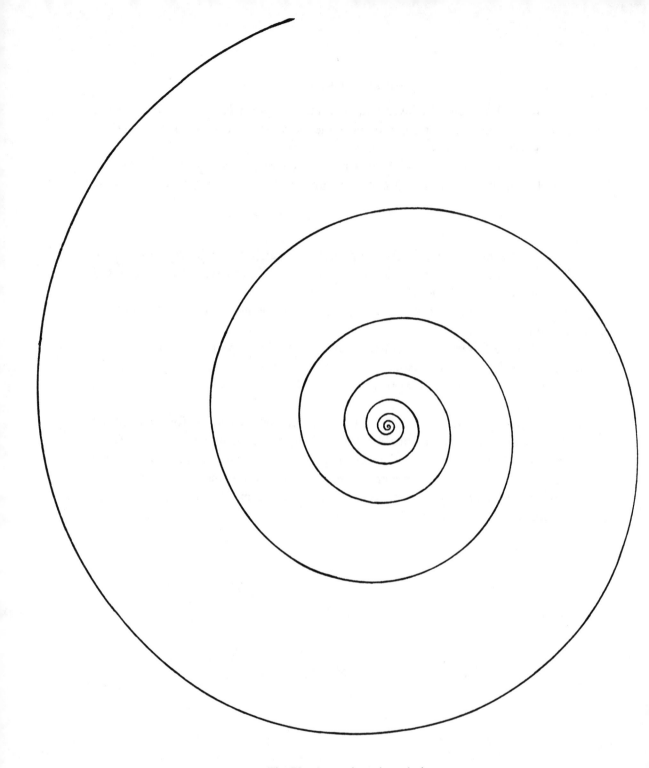

Fig. 70. An equiangular spiral

11

THE EQUIANGULAR SPIRAL

The Problem of the Four Dogs

A courtyard *ABCD* is a square of side 200 ft. (Fig. 71). Four dogs are started simultaneously from the four corners, the one at *A* facing towards *B*, the one at *B* towards *C*, and so on. Each dog pursues the next at a uniform speed of 20 ft. per sec. If *A'*, *B'*, *C'* and *D'* are simultaneous positions of the dogs that started at

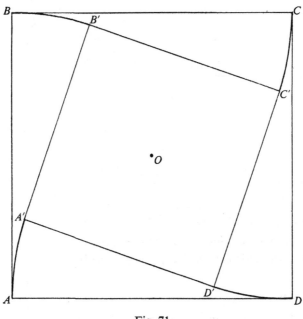

Fig. 71

A, *B*, *C* and *D* respectively, it is evident from symmetry that *A'B'C'D'* will be a square and the direction of motion of each dog will be along one side of the square, i.e. at 45° to the line joining it to the centre *O* of the courtyard. This will be true in any position and the fact characterizes the curve along which a dog moves. Such a curve, in which the tangent at any point makes a constant angle with the radius drawn to that point from a fixed point, is called an *equiangular spiral*.

As the dog at B' is moving at right angles to $A'B'$, the distance $A'B'$ is diminishing at 20 ft. per sec. and will be reduced to zero in 10 sec. from the start. The total length of the curve from A to the point where they meet is therefore 200 ft.

$A'B'$ is a tangent to the curve AA' and a normal to BB'. As this is so in any position, it follows that AA' is the evolute of BB' and the point A' is the centre of curvature for the curve BB' at B'. Thus each dog is at the centre of curvature of the path of the next one. (As the radius of curvature diminishes, the dogs will in practice slip, at a moment depending on the coefficient of friction.)

To Draw an Equiangular Spiral

Draw a series of lines, at equal intervals, radiating from a fixed point (called the *pole*). From a point on one of these lines draw a perpendicular to the next; from the foot of that perpendicular draw a perpendicular to the next line; and so on. A freehand curve may then be drawn through all the points so found.

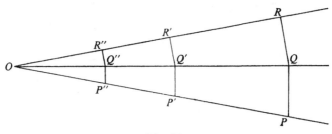

Fig. 72

Suitable Dimensions

The pole should be in the middle of the paper and the radiating lines may conveniently be taken at 10° intervals. The first point should be near the middle of a long edge of the paper. Several complete turns of the spiral should be drawn; it may also be continued backwards from the starting-point.

Geometrical Properties

Fig. 72 shows three successive points, P, Q, R, of the drawing described above, O being the pole. P', Q', R' are the three corresponding points of the next circuit, and P'', Q'', R'' those of the following circuit.

Triangles OPQ, OQR, ..., are similar, since corresponding angles are equal. Therefore, $OQ/OP = OR/OQ = ...$, and the lengths OP, OQ, OR, ..., are in geometrical progression. But the triangles $OP'Q'$, $OQ'R'$, ..., are part of the same series of triangles, and OP', OQ', OR', ..., are terms of the same geometrical progression. Again, OP, OP', OP'', ..., form a geometrical progression, for they are

[100]

equally spaced terms of the one just mentioned. (This provides a useful check on the accuracy of the drawing, especially in the part near the pole.)

Another geometrical progression is formed by the lengths PQ, QR, ..., $P'Q'$, $Q'R'$, ... Moreover, the quadrilaterals $PQQ'P'$, $QRR'Q'$, ..., are all similar, for they have corresponding angles equal and pairs of sides in proportion.

If the diagram had been drawn by marking lengths OP, OQ, OR, ..., in geometrical progression, along equally spaced radii, the same similarity properties would have held good, though the angles OQP, ORQ, ..., would not necessarily have been right angles.

Intermediate Points on the Curve: Polar Equation

The method given above for drawing the curve determines certain points P, Q, R, but not the intermediate points. To preserve the similarity properties, i.e. to make them applicable at all points on the curve, it is only necessary to ensure that the radii drawn at any equal intervals of angle should be in geometrical progression. The problem of determining intermediate points therefore reduces to that of inserting geometric means between the lengths OP, OQ, ..., of existing radii. Thus, for example, the radius bisecting angle POQ should be of length $\sqrt{(OP.OQ)}$.

More generally, let POQ be taken as a unit of angle, and let $OQ/OP = OR/OQ = k$. Then, if $OP = r_0$, OQ will be $r_0 k$ and OR will be $r_0 k^2$. At a point three units of angle from OP the radius will be $r_0 k^3$, and so on. On this principle we may define intermediate points (and indeed all points) on the curve by the equation

$$r = r_0 k^\theta,$$

where r is the length of the radius making an angle of θ units with the initial radius OP, of length r_0. This is the polar equation of the curve. (See also p. 107 for a more usual form of the equation.)

The constant k in this equation may be greater or less than 1. In the method of drawing suggested above, it is less than 1 and the radii OP, OQ, OR, ..., diminish as θ increases; but, if it is greater than 1, r increases with θ and in fact, if k is equal to the reciprocal of its former value, an identical curve is produced (or, to be precise, a mirror image of the former curve).

It should also be noted that the ratio k depends on the unit of angle chosen. Thus in the original drawing, if the unit is 10^c, $k = \cos 10^\circ = 0.9848$. If the unit of angle is changed to one revolution, the new value of k is $(0.9848)^{36}$.

The Equiangular Property

If now OU, OV, OW, ..., are equally spaced radii at any part of the curve (Fig. 73), so that the values of θ, namely θ_1, θ_2, θ_3, ..., are in arithmetical progression, it follows from the above equation that the values of r, namely r_1, r_2, r_3, ..., are in geometrical progression; for $r_2/r_1 = k^{\theta_2 - \theta_1} = k^{\theta_3 - \theta_2} = r_3/r_2$. Hence the triangles OUV, OVW, ..., are similar. If now U, V, W, ..., are taken very close together, the angles OUV, OVW, ..., will be very nearly the angles between the tangents to the curve and the radii OU, OV, ... But angles OUV, OVW, ..., are all equal. Hence, in the limit, the angle between the tangent to the curve and the radius is constant.

Similarity Properties

The similarity properties mentioned on p. 100 for the plotted points P, Q, R, ..., P', Q', R', ..., are now seen to be applicable at any points of the curve U, V, W, ..., provided always that they are spaced at equal intervals of angle. These facts account for many of the remarkable properties of the curve. If any part of the

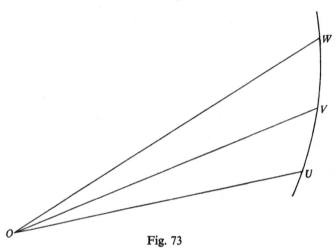

Fig. 73

curve is enlarged or reduced in any ratio, it becomes congruent to another part; and, if the whole curve from any point to the pole is enlarged or reduced, it becomes congruent to the same curve from another point to the pole. This can be illustrated by rotating the curve about the pole as centre. (A pin or spike may be pushed through the paper at the pole and the paper rotated in a vertical plane.)

Certain shells and fossils have forms closely resembling the equiangular spiral, the animal living in the shell having grown uniformly and having occupied successive portions each similar to the last. (See the frontispiece.) Horns, nails and hairs tend to grow in this shape, a fact which may be explained by the two sides having rates of growth which are in a constant ratio.

An Unending Curve...

It appears from the method of drawing the curve that there is no point from which a further point cannot be found, whether we are proceeding in the outward or the inward direction. There are thus, in this sense, no end-points. Again, from the equation, there is a value of r for every value of θ, positive or negative; moreover r increases or decreases as θ changes, approaching zero in one direction and increasing indefinitely in the other. Thus we may say that, starting from any point, the curve makes an infinite number of circuits round the pole on either side of that point.

...of Finite Length

Imagine that the curve rolls (without slipping) along a straight line, the straight line being always a tangent to it. Let the point of contact at any moment be I, and let the radius OI make a constant angle α with the tangent (Fig. 74). Then I is the instantaneous centre, and the pole O will move at right angles to OI, i.e. in a

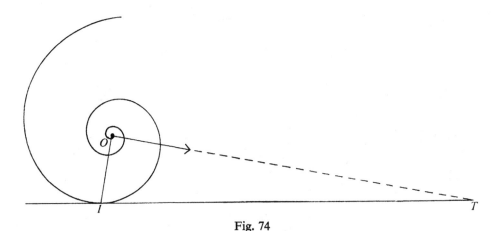

Fig. 74

fixed direction. It will, therefore, move in a straight line, meeting the fixed line at T, where $IT = OI\sec\alpha$. It follows that the length of the spiral from I to the pole is $OI\sec\alpha$.

** This paradox, that an unending curve should have a finite length,† depends on the precise meaning which we attach to the word *end*. Although there is no point of the curve itself which can be described as the end-point, nevertheless there is a point, namely the pole, which the curve approaches and beyond which it does not go. If we divide the curve by successive units of angle and the length of one part is l, the succeeding parts will be lk, lk^2, lk^3, ..., and the whole length to the pole may be defined as the limiting sum of the series

$$l + lk + lk^2 + lk^3 + \ldots$$

† Wallis wrote of this curve (*Opera*, vol. I, p. 561, 1695) 'Habes itaque *curvam interminabilem terminatae rectae aequalum*'.

Since k is less than 1 (for we are proceeding towards the pole), this is a convergent geometric series. It will be noticed that we have divided the curve into diminishing sections of arc-length not unlike the diminishing intervals of time in the better-known paradox of *Achilles and the Tortoise*.

Caustic

If a ray of light from a source at O is reflected by the curve at P, the envelope of the reflected ray, as P varies, is an equal spiral.

Proof: In Fig. 75, let R be the image of O in PN. Then PR is the reflected ray.

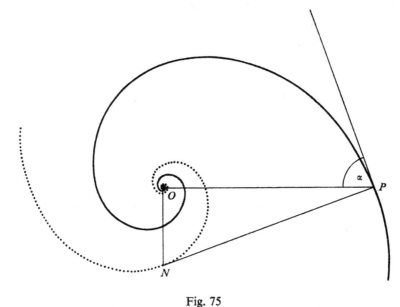

Fig. 75

Since $OR = 2OP\cos\alpha$ and the angle POR is constant, the locus of R is a spiral similar to, and therefore equal to, the original spiral. But angle $ORP = \alpha$; therefore RP is a tangent to this spiral, and the result follows.

A Roulette

Suppose that, in Fig. 75, the tangent at P meets NO produced at T. Then $PT = OP\sec\alpha$, and this is the arc-length of the curve from P to the pole. Therefore, if the tangent PT rolls on the curve without slipping, T is a fixed point relative to it. But OT makes a constant angle with OP and its length is proportional to OP; therefore the locus of T is a spiral similar to, and hence equal to, the original spiral.

[104]

Evolute

Let P be any point on the spiral, with α the constant angle between OP and the tangent. Let PN be the normal, meeting at N a line through O at right angles to OP (Fig. 75). Then angle $ONP = \alpha$. As ON is in a fixed ratio to OP and makes a constant angle with it, the locus of N is a spiral similar to, and therefore equal to, the original spiral. Since angle $ONP = \alpha$, PN is a tangent to this new spiral. But it is also a normal to the original curve; therefore the new spiral is the evolute of the original one, and N is the centre of curvature for the original spiral at P.

As shown on p. 84 for the cycloid, the arc of the evolute is equal to the difference of the radii of curvature at its end-points. The length of the new spiral from N to the pole (or as near to the pole as may be) is, therefore, equal to the difference between NP and a quantity which may be as near to zero as we please. Thus we may say that the length from N to O is NP, or $ON \sec \alpha$. This agrees with the result obtained above.

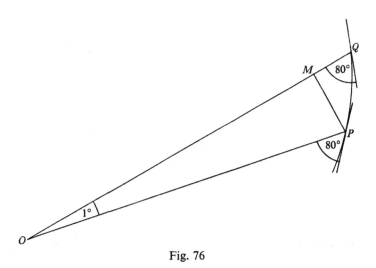

Fig. 76

Relation Between the Ratio k and Constant Angle

It is evident that there must be a relation between k and α, since either of these quantities determine the shape of the curve. In particular cases it is easy to calculate k approximately for any given value of α. For example, if $\alpha = 80°$, suppose that OP and OQ are radii $1°$ apart, with OQ greater than OP (Fig. 76). Let PM be the perpendicular from P to OQ. Then $PM = OP \sin 1°$, and QM is approximately $OP \sin 1° \cot 80° = OP \times 0.003077$. But $OM = OP \cos 1°$, $= OP$ approx.

$$\therefore \quad OQ = 1.003077 \times OP \text{ approx.} \quad \text{and} \quad OQ/OP = 1.003077.$$

If the unit of angle is $10°$, $k = (1\cdot003077)^{10} = 1\cdot033$. If it is one revolution, $k = (1\cdot003077)^{360} = 3\cdot022$. (For reasonable accuracy in these results it is necessary to use at least five-figure tables for the last stage of the calculation.)

The following table gives values of k and $1/k$, for a unit of one revolution, for a number of values of α:

α	k	$1/k$
89°	1·17	0·896
88°	1·25	0·803
85°	1·73	0·577
80°	3·03	0·330
75°	5·38	0·186
70°	9·84	0·102
60°	37·6	0·0266
45°	535	0·00187
30°	53250	0·0000188

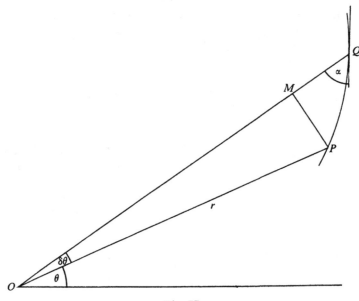

Fig. 77

** To express this relationship algebraically it is necessary to use calculus. We draw the figure (Fig. 77) with r increasing as θ increases: k will then be greater than 1. θ is now measured in radians. PM is drawn at right angles to OQ, angle POQ being $\delta\theta$. $OP = r$ and $OQ = r + \delta r$. Then $MQ = \delta r$ approx., and $PM = r\,\delta\theta$ approx. Angle $PQM = \alpha$ approx., and, in the limit, $\cot\alpha = dr/(r\,d\theta)$. Integrating with respect to θ, $$\log_e r = \theta\cot\alpha + \text{const.}$$

If the initial value of r (i.e. the value when $\theta = 0$) is r_0, the constant of integration is $\log_e r_0$. Hence $$\log_e r = \theta\cot\alpha + \log_e r_0,$$

[106]

which may be written as $\qquad \log_e(r/r_0) = \theta \cot \alpha,$

or $\qquad\qquad\qquad\qquad r = r_0 e^{\theta \cot \alpha}.$

This last is the usual form of the polar equation of the equiangular spiral. The curve is also called the *logarithmic spiral* because of the logarithmic form of the equation.

In these equations, θ is measured in radians. To obtain the values of k mentioned in earlier paragraphs it is only necessary to substitute for θ the value, in radians, of whatever is to be taken as the 'unit of angle'. Thus for the table given on p. 106, we substitute 2π for θ. Then $k = r/r_0 = e^{2\pi \cot \alpha}$.

Further Drawing Exercises

1. On radii drawn at $10°$ intervals mark distances from the pole proportional to the antilogarithms of $0.00, 0.01, 0.02, ...$ (This is best done on polar graph paper, the first point being plotted at a distance 10 cm. from the pole. When the curve has been taken as far as possible in the outward direction, it may then be drawn inwards as well by using the antilogarithms of $\bar{1}.99, \bar{1}.98, \bar{1}.97, ...$ If ordinary paper is used, the same scale is suitable.)

Calculate the value of k for one revolution of this spiral, and also the size of the constant angle α. Check by measurement.

2. Draw the pedal or negative pedal of one of your spirals with respect to its pole. (See pp. 153, 157).

Prove that the pedal or negative pedal of an equiangular spiral is an equal spiral.

3. Use dividers to measure radii in various directions on an ammonite or similar fossil or shell. Draw a graph showing the logarithm of the radius as a function of the angle.

If no suitable fossil or shell is available, the frontispiece of this book may be used, or the figures in the following table:

θ (degrees)	r (mm.)
0	8.1
45	9.0
90	10.3
135	11.4
180	13.0
225	15.0
270	17.2
315	20.0
360	22.1
405	24.5
450	27.0

Calculate the constant angle α for this spiral. (They are often about $85°$.)

4. Using radii at 30° intervals draw an equiangular spiral having a ratio k equal to 2 for a unit of one revolution. (*Hint:* The ratio for a unit of 30° will be $2^{1/12}$ and its logarithm will be $\frac{1}{12}\log 2$, i.e. 0·0251. From this, with the aid of anti-logarithms, a table can be quickly compiled giving the lengths of twelve successive radii, starting from any arbitrary value. For a second circuit it is only necessary to double these values.) Label the successive radii according to the notes of a musical scale, at semitone intervals, the 'higher' notes being towards the centre of the spiral. The lengths of the radii will then be proportional to the lengths of a vibrating string required to produce the corresponding notes (in equal-temperament tuning), each circuit of the spiral representing one octave. If the labelling is done in the opposite direction, the lengths of the radii will be proportional to the frequencies of the notes.

5. Plot the values used in no. 4 (or the lengths of successive radii of any other equiangular spiral) as equally-spaced ordinates of an ordinary graph. The resulting curve is the *exponential curve* and its chief property is that the gradient at any point is proportional to the ordinate. Such curves are of considerable importance in physics, illustrating for example the rate at which radium decays or the rate at which an alternating current dies away when switched off. They can also be used to represent the amount of a sum of money increasing at compound interest or the 'natural' growth of a population. An approximate method for drawing these curves without using tables is given on p. 122.

The Equiangular (or Logarithmic) Spiral: Summary

** 1. The polar equation is $r = ae^{\theta \cot \alpha}$.

2. $\phi = \alpha$. (The curve makes a constant angle with the radius vector.)

3. A line through the pole meets the curve at distances in geometric progression, the common ratio being as given in the table on p. 106.

4. The pedal equation is $p = r \sin \alpha$.

5. The length of the arc (measured from the pole) is $r \sec \alpha$.

6. $\rho = r \operatorname{cosec} \alpha$.

7. The evolute is an equal spiral.

8. The inverse with respect to the pole is an equal spiral.

9. The pedal and negative pedals with respect to the pole are spirals equal to the original curve.

10. The caustic, with radiant point at the pole, is an equal spiral.

11. If the curve is rolled along a fixed straight line, the locus of the pole is a straight line.

12. If any part of the curve is enlarged or reduced in any ratio, it becomes congruent to another part of the same spiral.

The equiangular spiral was first considered in 1638 by Descartes, who started from the property $s = a.r$. Torricelli, who died in 1647, worked on it independently and used for a definition the fact that the radii are in geometric progression if the angles increase uniformly. From this he discovered the relation $s = a.r$; that is to say, he found the rectification of the curve. John Bernoulli, some fifty years later, found all the 'reproductive' properties of the curve (as, for example, nos. 7, 8, 9, 10, 12, above); and these almost mystic properties of the 'wonderful' spiral made him wish to have the curve incised on his tomb: *Eadem mutata resurgo*—'Though changed I rise unchanged'.

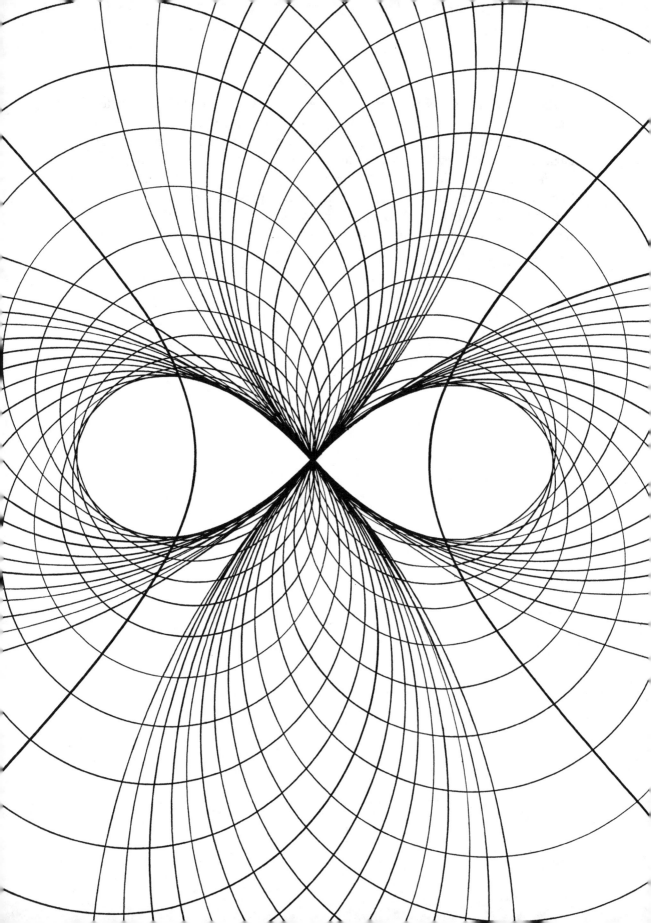

12

THE LEMNISCATE OF BERNOULLI

To Draw a Lemniscate (First Method)

Draw a circle, centre C, and mark a point O whose distance from C is $\sqrt{2}$ times the radius (Fig. 79). Through O draw any straight line cutting the circle at Q, Q'. On this line mark points P and P' at distances from O equal to QQ'. Repeat for many positions of the line. The locus of P and P' is the lemniscate.

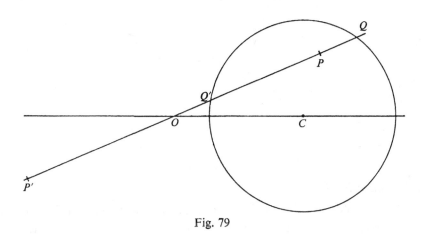

Fig. 79

Suitable Dimensions

The circle should be near the middle of the right-hand edge of the paper, with O to the left of it.

		Radius ($= \frac{1}{2}a$)		OC ($= \frac{1}{2}a\sqrt{2}$)	
Paper:	1_L	1·5 in.	or 4 cm.	2·12 in.	or 5·66 cm.
	2_L	2 in.	5 cm.	2·83 in.	7·07 cm.
	3_L	2·5 in.	6 cm.	3·54 in.	8·48 cm.

To Draw a Lemnsicate (Second Method)

In Fig. 80, with O and C as before, OC is produced to S, so that $OC = CS = \frac{1}{2}a\sqrt{2}$. A line is drawn through S parallel to QO, and perpendiculars PR, ON, $P'R'$ are drawn to it from P, O, P'. CM is the perpendicular from C to OQ. Then

$$NR = OP = Q'Q = 2Q'M \quad \text{and} \quad ON = 2CM.$$

[111]

Fig. 78. The rectangular hyperbola and the lemniscate of Bernoulli

Therefore, the right-angled triangles ONR, CMQ' are similar, and

$$OR = 2CQ' = a.$$

Hence R, and similarly R', lie on a circle whose centre is O and whose radius is a.

The lemniscate may therefore be drawn as follows:

Draw a circle, centre O, and mark a point S whose distance from O is $\sqrt{2}$ times the radius. Place two set squares as shown in Fig. 81, so that OP and RS are parallel. If R lies on the circle, the locus of P will be the lemniscate.

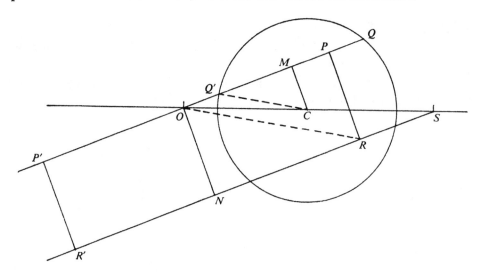

Fig. 80

Suitable Dimensions

The circle should be in the middle of the paper and its radius (a) should be double that of the circle used in the first method. OS should be double the length given for OC.

The Lemniscate as a Pedal Curve

The line PR (Fig. 81) touches a rectangular hyperbola whose centre is O, S being one focus. (See p. 25.) P is thus the foot of the perpendicular from the centre of the hyperbola to a tangent. The lemniscate is, therefore, the pedal of a rectangular hyperbola with respect to its centre.

To Draw a Lemniscate (Third Method)

Draw a rectangular hyperbola and let its centre be O. With any point T on the hyperbola as centre and with radius TO draw a circle. Repeat many times for different positions of T. As P varies, the envelope of this circle is the lemniscate (Fig. 78).

[112]

Suitable dimensions: The base-circle of the hyperbola should have radius as given for the First Method, the distance from the centre O to the focus S being that given for OC.

Proof: If the circles whose centres are at points T, t on the hyperbola intersect at P'' (Fig. 82), triangles OTt and $P''Tt$ are congruent and P'' is the image of O in the chord tT. As t approaches T, the chord becomes a tangent and P'' becomes a

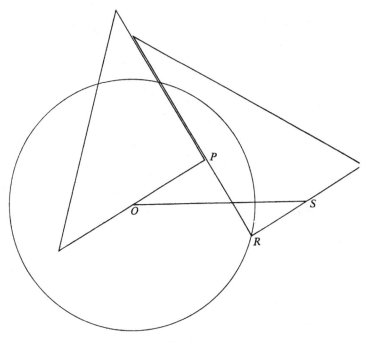

Fig. 81

point on the envelope. If OP'' cuts the tangent at P, the locus of P is a lemniscate (being the pedal of a rectangular hyperbola with respect to its centre), and that of P'' is a similar curve on double scale.

Polar Equation of the Lemniscate

In Fig. 80, let OP be r and let angle SOP be θ. Then $ON = OS\sin\theta = a\sqrt{2}\sin\theta$, and

$$r^2 = OP^2 = NR^2 = OR^2 - ON^2$$
$$= a^2 - 2a^2\sin^2\theta$$
$$= a^2\cos 2\theta.$$

This is the polar equation of the lemniscate.

Exercises

1. *To draw a lemniscate (fourth method).* Draw two circles of radius a, their centres C, C' being a distance $a\sqrt{2}$ apart. Find points M, M', one on each circle, such that $MM' = a\sqrt{2}$, and MM' is not parallel to CC' (Fig. 83). Mark the mid-

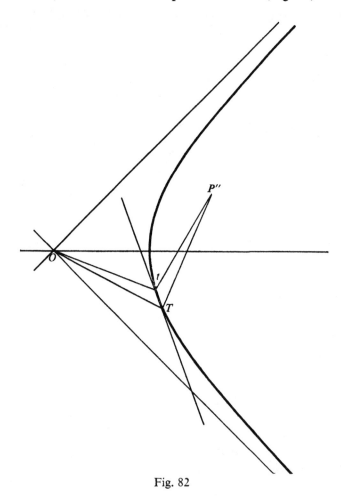

Fig. 82

point P of MM'. The locus of P is a lemniscate. (This method is that of the linkage shown in Fig. 84.)

Suitable Dimensions

		$CC'\ (= a\sqrt{2})$		Radius $CM\ (= a)$	
Paper:	1_L	3 in. or	8 cm.	2·12 in. or	5·66 cm.
	2_L	5 in.	12 cm.	3·54 in.	8·48 cm.
	3_L	6 in.	14 cm.	4·24 in.	9·9 cm.

2. *The lemniscate as the inverse of a rectangular hyperbola.* Draw a rectangular hyperbola with centre O (see p. 26). Draw a straight line through O, meeting the hyperbola at Q, and mark on it a point P such that $OP.OQ = k^2$, k being a constant. Then the locus of P is a lemniscate. (This may be proved from the polar equation of the rectangular hyperbola $r^2\cos 2\theta = a^2$. See p. 32.) The points P may be determined either by using a table of reciprocals or by the following construction:

Draw any convenient circle not passing through O. On this circle find a point F such that $OF = OQ$. Let OF cut the circle again at E. Then $OP = OE$.

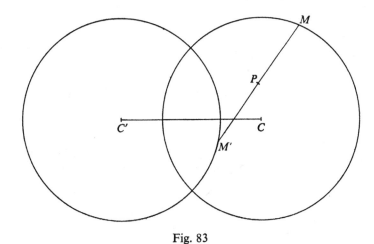

Fig. 83

Suitable dimensions: If a table of reciprocals is used, it is convenient for k^2 to be 10 sq. units. The semi-axis of the hyperbola may then be 2 cm. (on small paper) or 2 in. (on larger paper).

If the geometrical construction is used, the hyperbola may be any size; the circle may conveniently be drawn to occupy most of the space between O and one corner of the paper.

* 3. Prove that, in Fig. 79, triangles OCQ, CPQ are similar.

(*Hint:* $QP.QO = OQ'.OQ = $ square on tangent from $O = QC^2$.)

Hence prove that $CP/OC = CQ/OQ$, and similarly that $CP'/OC = CQ'/OQ'$. Use these facts to prove that $CP.CP' = OC^2$, and hence that, if C' is a point on CO produced, such that $CO = OC'$, then $PC.PC'$ is constant. (The lemniscate may thus be defined as the locus of a point the product of whose distances from two fixed points is equal to the square of half the distance between them.)

[115]

** 4. In Fig. 83, use Apollonius' Theorem to prove that $CM'^2 = 2CP^2$ and $C'M^2 = 2C'P^2$; then use Ptolemy's Theorem to prove that

$$CM'.C'M = a^2.$$

Hence prove that $CP.C'P = \frac{1}{2}a^2 = OC^2$.

The Lemniscate: Summary

1. The polar equation (with pole at the centre) is $r^2 = a^2\cos 2\theta$.
2. The Cartesian equation is $(x^2+y^2)^2 = a^2(x^2-y^2)$.
3. The bipolar equation is $rr' = \frac{1}{2}a^2$.

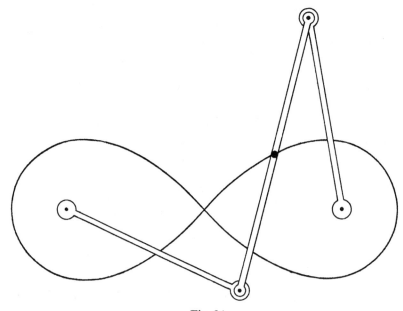

Fig. 84

4. The pedal equation is $r^3 = a^2p$.
5. $\phi = 2\theta + \frac{1}{2}\pi$.
6. $\psi = 3\theta + \frac{1}{2}\pi$.
7. $A = a^2$.
8. $\rho = \pm a^2/(3r)$.
9. The lemniscate is the cissoid of a circle with respect to a point whose distance from the centre is $\sqrt{2}$ times the radius.
10. It is the pedal of a rectangular hyperbola with respect to its centre.
11. It is the inverse of a rectangular hyperbola with respect to its centre.

[116]

In 1694 James Bernoulli published an article in the *Acta Eruditorum* on a curve 'shaped like a figure 8, or a knot, or bow of a ribbon', for which he used the Latin word *lemniscus* ('a pendent ribbon, fastened to a victor's garland'). He was not aware that this curve was a special case of a Cassinian Oval (see p. 187), and he did not investigate its geometrical properties; the main interest was analytical, and the investigations on the length of arc of the curve laid the foundations of the later work on elliptic functions.

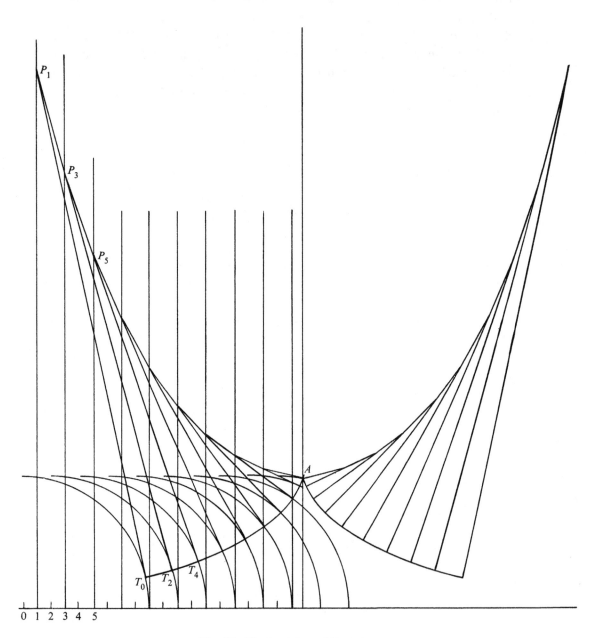

Fig. 85. The tractrix and the catenary

13

THE TRACTRIX AND CATENARY

To Draw the Tractrix and Catenary

Draw a base-line across the foot of the paper and mark points on it at equal intervals, beginning at the left-hand margin and continuing as far as the middle of the paper. Number these points 0, 1, 2, 3, ..., from left to right. Through points 1, 3, 5, ..., drawn lines at right angles to the base-line. (These will be called 'vertical lines 1, 3, 5, ...'.) With points 0, 2, 4, ..., as centres draw quadrants of circles of fixed radius c, as shown in Fig. 85. (These will be called 'quadrants 0, 2, 4, ...'.) On the vertical line 1 choose a point P_1 and from it draw tangents $P_1 T_0$ and $P_1 T_2$ to quadrants 0 and 2 respectively. With P_1 as centre, draw an arc from T_0 to T_2. Let $P_1 T_2$ cut the vertical line 3 at P_3. With centre P_3 draw an arc from T_2 to meet quadrant 4 at T_4. Joint $P_3 T_4$, cutting vertical line 5 at P_5; and so on.

The arcs joining $T_0, T_2, T_4, ...$, will form a curve approximating to a tractrix, and the points $P_1, P_3, P_5, ...$, will lie approximately on a catenary.

Suitable Dimensions

The paper may be placed either way. The intervals between the points 0, 1, 2, ..., may conveniently be 0·2 in. or 0·5 cm. The point P_1 should be as far from the base-line as possible. The point T_0 must be accurately determined, either by using the Euclidean construction for the first tangent or by drawing a perpendicular from the centre 0. If $P_1 T_0$ is beyond the stretch of the compasses, $P_1 T_2$ should be made equal to $P_1 T_0$ and $T_0 T_2$ joined by a straight line or a freehand curve.

When the tractrix has been continued until it meets the catenary (at a point A distant c from the base-line) the vertex of both curves has been reached. A line drawn through this point at right angles to the base-line is an axis of symmetry for both curves and should be used as such for drawing the other half of each of them.

Geometrical Properties

If the complete circles, of which quadrants 0 and 2 are parts, were drawn, they would intersect at points R and S on the vertical line 1. Then $P_1 T_2^2 = P_1 R . P_1 S$ (tangent and secant theorem) $= P_1 T_0^2$. Therefore, $P_1 T_2 = P_1 T_0$. By the same theorem, $P_3 T_4$ is a tangent to quadrant 4; and so on. Therefore the curve

$T_0 T_2 T_4$ ..., as drawn, crosses each of the quadrants at right angles. The *tractrix* is in fact the curve which cuts at right angles all the circles, of constant radius, whose centres lie on a straight line. The above method of drawing the curve will be the more accurate as the intervals between the points 0, 1, 2, ..., are made smaller.

The lines $P_1 T_0$, $P_1 T_2$, $P_3 T_4$, ..., are normals to the curve as drawn; and in the limit, as the intervals between the points 0, 1, 2, ..., are made smaller, they become

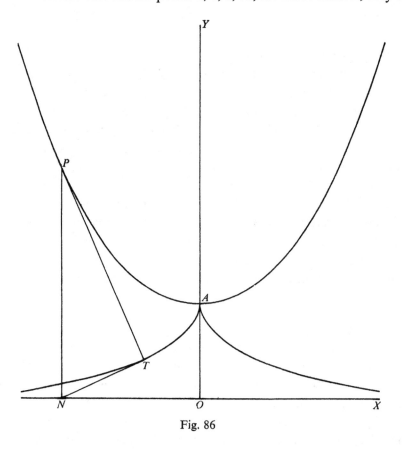

Fig. 86

normals to the tractrix and tangents to the catenary. The *catenary* is in fact the evolute of the tractrix. In Fig. 86, PT is a normal to the tractrix and a tangent to the catenary; PN is drawn perpendicular to the base-line and angle PTN is a right angle.

Length of arc of the catenary. In Fig. 85, the sum of the lengths $P_1 P_3$, $P_3 P_5$, ..., as far as the vertex, is equal to $P_1 T_0$. In the limit, PT (Fig. 86) is equal to s, the length of the arc PA. If the base-line is taken as axis of x, and the axis of symmetry as axis of y, $PN^2 = PT^2 + NT^2$, i.e. $y^2 = s^2 + c^2$. The length of the arc from P to the vertex is, therefore, $\sqrt{(y^2 - c^2)}$.

[120]

Gradient. The gradient, $\tan\psi$, of the catenary at P is equal to $\tan PNT = s/c$, s being taken as positive when x is positive.

Intrinsic equation. The equation $s = c\tan\psi$ is known as the intrinsic equation of the catenary.

Experiments

1. *The tractrix.* Attach a small object to the end of a string whose length is a convenient multiple of the length c in your drawing. Place the object on a horizontal table in such a position that the string just reaches to one edge of the table. Move the free end of the string along that edge and mark with chalk the path which the object follows. Compare the curve so formed with the tractrix as previously drawn.

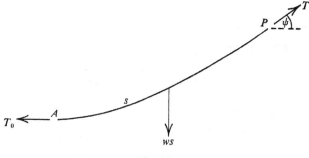

Fig. 87

Proof: If T is the object and N is the free end of the string (Fig. 86), the object always moves in the direction TN, i.e. at right angles to a circle whose centre is N and whose radius is the fixed length NT. The locus of T is therefore a tractrix as previously defined. (It is from this property that the name *tractrix* is derived.)

2. *The catenary.* Measure, in your drawing, the distance x from P_1 to the axis of symmetry; also the distance s from P_1 to T_0. Take a uniform flexible chain whose length is a convenient multiple of $2s$, and hang it freely with its ends on the same level at a distance apart which is the same multiple of $2x$. Measure the width at different heights and compare with your drawing.

** *Proof:* Let the weight of the chain be w per unit length. Let T be the tension at any point P where the gradient is ψ and the arc-length from the vertex is s (Fig. 87). Let the tension at the vertex be T_0. Considering the equilibrium of the portion of chain from P to the vertex, we have

$$\text{(resolving horizontally)} \quad T\cos\psi = T_0,$$
$$\text{(resolving vertically)} \quad T\sin\psi = ws.$$
$$\therefore \quad \tan\psi = ws/T_0.$$

[121]

This is the intrinsic equation of the catenary, and may be written in the form $s = c\tan\psi$, where $c = T_0/w$.

3. *Comparison of the catenary and the parabola.* Near the vertex, the catenary is indistinguishable from a parabola, the focus of the parabola being a point S on OY such that $OA = 2AS$. Use the method described on p. 3 to draw this parabola.

It is to be noted that the parabola is the curve formed by the chains of a suspension bridge, where light cables support a uniform heavy roadway. The distribution of weight is then uniform per horizontal unit of length. In the catenary formed by a hanging chain the distribution of weight is uniform per unit of arc-length. In large suspension bridges the chains may weigh as much as the roadway; the curve they form is then something between a catenary and a parabola.

Cartesian Equation of the Catenary

**At any point of the catenary

$$\frac{dx}{dy} = \cot\psi = \frac{c}{s} = \frac{c}{\sqrt{(y^2 - c^2)}}.$$

Integrating with respect to y,

$$x = c\,\mathrm{ch}^{-1}(y/c), \quad \text{i.e. } y = c\,\mathrm{ch}(x/c).$$

Further Drawing Exercises

1. *The catenary from tables.* Use a table of hyperbolic cosines to plot a catenary, taking the same value of c as in your original drawing. Compare the results of the two methods.

2. *The exponential curve.* Draw a base-line across the foot of the paper and mark points on it at equal intervals k. Through each of these points draw a line at right angles to the base-line. (These lines will be called 'vertical lines'.) Choose any one of the marked points as origin and label it '0', the points to the right of it being called 1, 2, 3, ..., and those to the left -1, -2, -3, ... On the vertical line 0 mark a point P_0 whose distance from the point 0 is $2 \cdot 5k$. Join the point -2 to P_0 and produce this line to meet vertical line 1 at P_1. Join the point -1 to P_1 and produce to meet vertical line 2 at P_2. Join the point 0 to P_2 and produce to meet vertical line 3 at P_3; and so on. The points $P_0, P_1, P_2, \ldots$, will lie on an exponential curve. Points to the left of the origin may be found by joining P_0 to the point -3, cutting vertical line -1 at P_{-1}; and so on.

Suitable dimensions. A sheet of ordinary ruled paper may be used, in the 'landscape' position, with the ruled lines serving as 'vertical lines'. The point 0 should be taken rather to the right of the middle of the paper. k may be 0·4 in.

[122]

Check by use of tables: Take the base-line as x-axis and the vertical line 0 as y-axis, with OP_0 (i.e. $2 \cdot 5k$) as unit of measurement on both axes. The equation of the curve as drawn is then $y = (2 \cdot 756)^x$. If the curve $y = e^x$ is plotted from tables, it will lie close alongside.

3. *The catenary in relation to the exponential curve.* Draw an exponential curve $y = e^x$ and also its mirror image in the y-axis ($y = e^{-x}$). On each vertical line mark the mid-point of the segment cut off from that line by the two curves. These mid-points will lie on a catenary.

The Catenary: Summary

** 1. The intrinsic equation is $s = c \tan \psi$.

2. The Cartesian equation is $y = c\,\mathrm{ch}(x/c)$.

3. $s = c\,\mathrm{sh}(x/c)$.

4. $y^2 = s^2 + c^2$.

5. $\rho = c \sec^2 \psi = \pm y^2/c$.

6. The catenary is the form assumed by a uniform flexible chain hanging freely under gravity.

7. It is the locus of the focus of a parabola which rolls on a straight line.

8. It is the evolute of a tractrix.

The Tractrix: Summary

** (x', y', s', ρ', ψ' refer to a point on the tractrix, the plain letters being used for the corresponding point of the catenary.)

1. Parametric equations are
$$x' = c \log\,(\sec \psi + \tan \psi) - c \sin \psi,$$
$$y' = c \cos \psi.$$

2. The Cartesian equation is $\pm x' = c\,\mathrm{ch}^{-1}(c/y') - \sqrt{(c^2 - y'^2)}$.

3. $\psi' = \psi \pm 90°$ (the $+$ or $-$ sign being taken according to whether x' is $-$ or $+$).

4. $\rho' = |s| = |c \tan \psi| = |c \cot \psi'|$.

5. $s' = -c \log \sin |\psi'| = c \log(c/y')$.

6. The evolute is a catenary.

7. The x-axis is an asymptote.

8. The length of the tangent, from the point of contact to the asymptote, is constant.

9. The area between the curve and its asymptote is $\frac{1}{2}\pi a^2$. (This may be proved by considering that the area swept out by the tangent is $\int \frac{1}{2}a^2\,d\psi$.)

10. The tractrix is the orthogonal trajectory of a set of circles, of constant radius, whose centres lie on a straight line.

[123]

11. It is the path of an object dragged along a horizontal plane by a string of constant length when the other end of the string moves along a straight line in the plane.

12. It is an involute of the catenary.

Galileo's suggestion that a heavy rope would hang in the shape of a parabola was disproved by Jungius in 1669, but the true shape of the 'chain-curve', the catenary, was not found until 1690/91, when Huygens, Leibniz and John Bernoulli replied to a challenge by James Bernoulli. David Gregory, the Oxford professor, wrote a comprehensive treatise on the 'catenarian' in 1697. The name was first used by Huygens in a letter to Leibniz in 1690.

The problem of the tractrix (see no. 11, above) was proposed to Leibniz by a French doctor. Leibniz found and used the property no. 7; Huygens not only solved the immediate problem (and gave the curve its name), but he succeeded in generalizing it.

When the tractrix is rotated about its asymptote it generates a surface of constant negative curvature; this property enabled Beltrami (1868) to construct a model of hyperbolic non-euclidean geometry, in which a 'plane' is represented by this surface. Incidentally, if a is the length of the tangent (no. 8, above), the solid of revolution has volume and surface area equal respectively to the volume and surface area of a sphere of radius a.

** The tractrix has been proposed (by Schiele) as the ideal form for a bearing supporting a revolving shaft which exerts on it a considerable longitudinal thrust (i.e. a thrust parallel to the axis of rotation). Suppose that, in Fig. 86, the base-line represents the axis of rotation and the area between it and the left-hand portion of the tractrix represents a half-section of the end of the revolving shaft. It is desirable that the effect of wear on the bearing should be to shift the whole curve a small distance δl to the left. The movement in the normal direction at T will be $\delta l \sin \psi$, where ψ is the angle ONT. The work done in causing such wear on any small area δA of the bearing is then proportional to $\delta A . \delta l \sin \psi$. But this is equal to the work done against friction on that area, namely $\mu p . \delta A . 2\pi y$ per revolution, where μ is the coefficient of friction, p is the normal pressure (supposed uniform) and y is the distance of T from the base-line. It follows that, to obtain the desired distribution of wear, y should be proportional to $\sin \psi$, i.e. $y \operatorname{cosec} \psi$ should be constant, as in fact it is for the tractrix and for no other curve.

PART II

WAYS OF FINDING
NEW CURVES

Many of the curves discussed in the preceding chapters were drawn from
a base-line or a base-circle, and sometimes other curves such as the parabola or hyperbola were used. The strophoid, for example, was drawn first
by using a straight line and a point (the first and second methods); then
by using a parabola (the envelope method); and finally by using a circle
and its diameter (the 'cissoid' method). It has also been seen that any
curve has its evolute, and its pedal or negative pedal with respect to a given
point. Curves have been obtained too by rolling circles along lines or other
circles, and by sliding set squares against fixed points or along fixed lines.

These and other methods of obtaining new curves will now be considered. The reader will often be able to recall the use already made of
a method and will then be able to use it to draw new curves. There is no
end to the number and variety of plane curves: many are well known and
and have been studied in detail; others are known, while their geometrical
properties remain undiscovered; and there are many more which have yet
to be drawn for the first time.

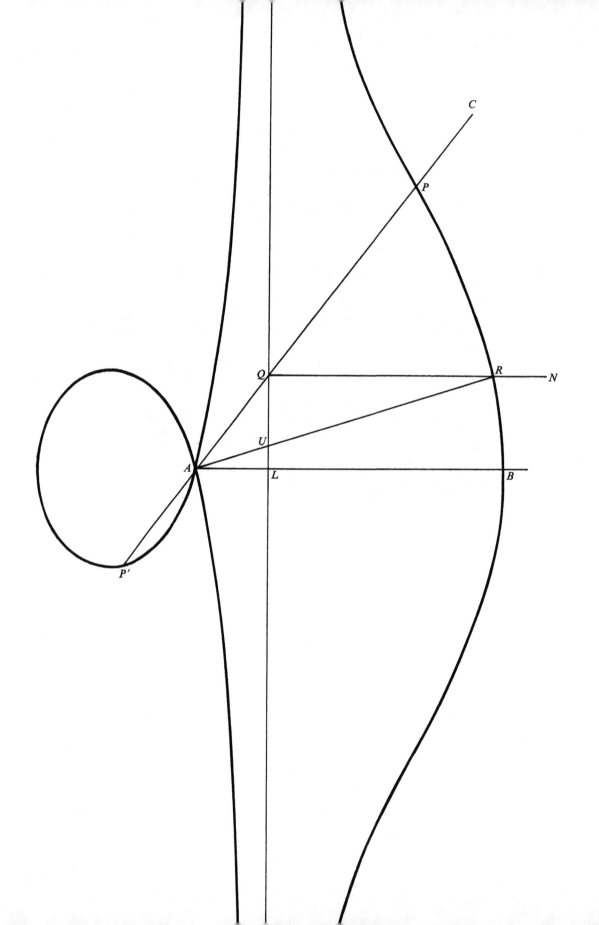

14

CONCHOIDS

Definition

Let S be any curve and let A be a fixed point. If a straight line is drawn through A to meet S at Q, and if P and P' are points on this line such that

$$P'Q = QP = k \text{ (constant)},$$

the locus of P and P' is called a *conchoid of S with respect to A*.

The conchoid of a circle with respect to a point on its circumference is the limaçon (p. 45). If the fixed distance is equal to the diameter of the circle, it is the cardioid (p. 36).

The Conchoid of Nicomedes

This is the conchoid of a straight line with respect to a point not on the line. Three cases arise, with the fixed distance less than, equal to or greater than the distance from the point to the line. The last of these is shown in Fig. 88, with A the fixed point and LQ the fixed straight line.

The polar equation, with A as pole and the perpendicular AL from A to the fixed line as initial line, is $r = a\sec\theta \pm k$, a being the length of AL.

This conchoid can be used for trisecting an angle, as follows: In Fig. 88, suppose that $AQ = \frac{1}{2}QP$. Let QR be drawn parallel to AL, meeting the outer branch of the conchoid at R; and let AR cut the fixed line at U. Then angle $RAL = \frac{1}{3}$ of angle QAL. (*Hint for proof:* Join Q to the mid-point of UR.) Hence the following method:

If BAC is the given angle, draw any line LQ at right angles to AB, cutting AC at Q. Draw QN parallel to LB. Let the conchoid of LQ with respect to A, with fixed distance equal to $2 \times AQ$, cut QN at R. Then angle $BAR = \frac{1}{3}$ of angle BAC.

(In practice it is not necessary to draw the conchoid, but only to place a ruler so that its edge passes through A and makes an intercept equal to $2 \times AQ$ between QL and QN.)

The curve may be drawn mechanically, by means of a ruler PP' which is made to

[127]

Fig. 88. A conchoid of Nicomedes and its use for trisecting an angle

pass through a fixed point A, its mid-point Q moving along a fixed line LQ. The ends P and P' then describe the curve. The instantaneous centre is the point where a line through A perpendicular to the ruler meets RQ produced, and the normals at P and P' must, therefore, pass through this point.

Other Conchoids

The following are suggested for drawing:

1. The conchoid of a circle of radius a with respect to a point within it at a distance b from the centre (i) with the fixed distance between $(a-b)$ and a, (ii) with the fixed distance equal to a.

2. The conchoid of a lemniscate with respect to its centre, the fixed distance being equal to the distance from the centre to the furthest point of the curve.

3. *The focal conchoids of the conic sections.* (In these conchoids the outer branch is of little interest, having roughly the same shape as the original conic; but the inner branch takes various forms according to the magnitude of the fixed distance. Thus, for the parabola, if the distance from the focus to the vertex is a, the form varies according to whether the fixed distance is (i) less than a, (ii) equal to a, (iii) between a and $2a$, (iv) equal to $2a$, or (v) greater than $2a$. For the ellipse, if $A'A$ is the major axis, with C the centre and S the focus nearer to A, and SL the semi-latus-rectum, the two principal forms occur (i) when the fixed distance is between SA and SL, or between CA and SA', (ii) when it is between SL and CA.)

The Conchoid of Nicomedes: Summary

** 1. The polar equation is $r = a\sec\theta + k$. (It is not necessary to write $\pm k$, since points on the inner branch are given by values of θ between $90°$ and $270°$.)

2. The Cartesian equation is $k^2x^2 = (a-x)^2(x^2+y^2)$. (When $k < a$, this equation includes an isolated point at the origin.)

3. The asymptote is $x = a$.

4. If $k > a$, there is a node at the origin;
   if $k = a$, there is a cusp;
   if $k < a$, there is neither.

5. The gradient at the node is given by $\cos\psi = \pm a/k$.

6. Area of loop

$$= a\sqrt{(k^2-a^2)} - 2ak\log\frac{k+\sqrt{(k^2-a^2)}}{a} + k^2\cos^{-1}\frac{a}{k}.$$

(The area between either branch and the asymptote is infinite.)

7. The normals at P and P' (Fig. 88) pass through the point where a line through A perpendicular to AP meets RQ produced.

The invention of the conchoid ('mussel-shell shape') is ascribed to Nicomedes (second century B.C.) by Pappus and other classical authors; it was a favourite with the mathematicians of the seventeenth century as a specimen for the new method of analytical geometry and calculus. It could be used (as was the purpose of its invention) to solve the two problems of doubling the cube and of trisecting an angle; and hence for every cubic or quartic problem. For this reason Newton suggested that it should be treated as a 'standard' curve.

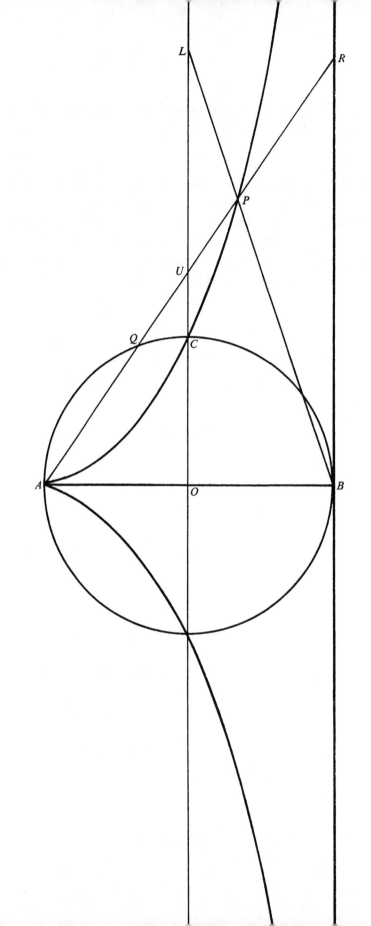

15

CISSOIDS

Definition

Let S and S' be any two curves and let A be a fixed point. A straight line is drawn through A cutting S and S' at Q and R respectively, and a point P is found in the line such that $AP = QR$, these lengths being measured in the direction indicated by the order of the letters. Then the locus of P is called the *cissoid of S and S' with respect to A*.

Thus the cissoid of two concentric circles, radii r_1, r_2, with respect to their common centre is a circle with the same centre and radius $|r_1 - r_2|$.

The Cissoid of Diocles

This is the cissoid of a circle and a straight line touching it, with respect to the point on the circumference of the circle diametrically opposite to the point of contact. In Fig. 89, A is the fixed point, S and S' are the circle and the tangent at B, and $AP = QR$.

This curve may be used for finding two mean proportionals between two given lengths. In Fig. 89, OU is the first of two mean proportionals between OC and OL; or, if the circle is of unit radius, the measure of OU is the cube root of that of OL. (*Hint for proof*: Let AO be a and let angle OAP be θ. Express coordinates of U, P and L in terms of a and θ.)

For other properties of this curve, see the summary given below.

Other Cissoids

1. *The oblique cissoid.* The cissoid of a circle and a straight line touching it, with respect to a point on the circumference not directly opposite the tangent, is again a curve having a cusp at the given point, with the given straight line as asymptote; but the curve now crosses the asymptote.

2. The cissoid of a circle and a straight line not a tangent, with respect to a point on the circumference, is a curve which has a node and loop if the straight line cuts the circle. If the straight line passes through the centre of the circle, it is a strophoid. (See p. 95.)

[131]

Fig. 89. The cissoid of Diocles and its use for finding a cube root

3. The cissoid of two intersecting straight lines with respect to a point not on either of them is a curve of one branch having asymptotes parallel to the given lines.

4. If a line through a fixed point A meets a curve S in two points Q and R, a cissoid of S with respect to A may be drawn. (The two curves S and S' are now replaced by two parts of the curve S.) Q and R are interchangeable, so the method is to mark on the variable line two points P and P' such that $P'A = AP = QR$. The curve thus obtained will have central symmetry about A. The lemniscate was drawn in this way (p. 111). The cissoid (in this sense) of a parabola with respect to a point outside it not on its axis is a double S-shaped curve; while that of a cardioid with respect to a point on its axis produced is a figure-of-eight curve with two cusps.

The Cissoid of Diocles: Summary

** 1. The polar equation (pole at A, initial line AB) is

$$r = 2a(\sec\theta - \cos\theta) \quad \text{or} \quad r = 2a\sin^2\theta/\cos\theta.$$

2. Parametric equations, using t for $\tan\theta$ as parameter, with A as origin and AB as axis of x, are

$$x = \frac{2at^2}{1+t^2}, \quad y = \frac{2at^3}{1+t^2}.$$

3. The Cartesian equation (with the same axes) is

$$y^2(2a-x) = x^3.$$

4. The asymptote is $x = 2a$.

5. The area between the curve and its asymptote is $3\pi a^2$.

6. The cissoid is the inverse of the parabola $y^2 = 8ax$ with respect to the origin (radius of inversion $4a$).

7. It is the pedal of the parabola $y^2 = -8ax$ with respect to the origin.

8. It is the negative pedal of a cardioid with respect to the point opposite the cusp-point. (See pp. 155, 157.)

9. It is the locus of the vertex of a parabola which rolls on an equal fixed parabola, starting from the position in which the vertices coincide.

10. If a set square moves as in Fig. 65 (p. 92), the locus of the mid-point of PQ is a cissoid.

The name *cissoid* ('ivy-shaped') is mentioned by Geminus in the first century B.C., that is, about a century after the death of the inventor Diocles. In the commentaries on the work by Archimedes *On the Sphere and the Cylinder* the curve is referred to as Diocles' contribution to the classic problem of doubling the cube. In Fig. 89, if $OC = a$ and $OL = 2a$ then $OU^3 = 2a^3$.

The mathematicians of the seventeenth century tried their skill on the cissoid. Fermat and Roberval constructed the tangent (1634); Huygens and Wallis found the area (1658); while Newton gives it as an example, in his *Arithmetica Universalis*, of the ancients' attempts at solving cubic problems and again as a specimen in his *Enumeratio Linearum Tertii Ordinis*.

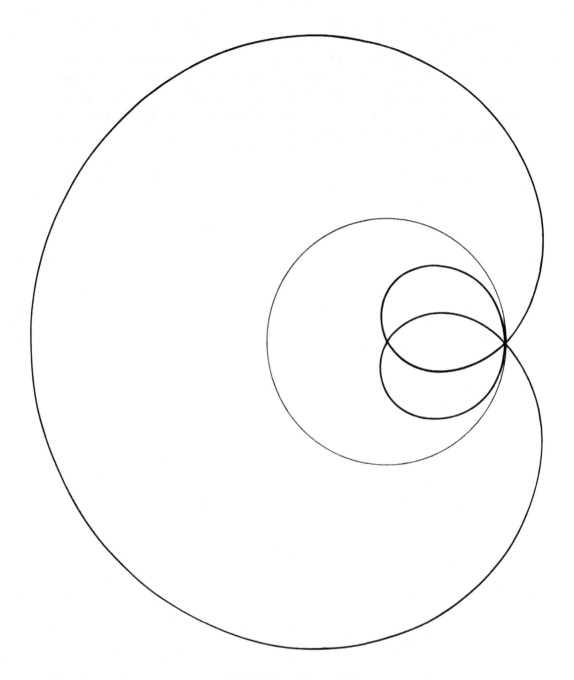

Fig. 90. Freeth's nephroid

16

STROPHOIDS

Definition

Let S be any curve and O a point (called 'the pole') and A another point (called 'the fixed point'). Then if a variable line through O meets the curve S at Q, and points P, P' are marked on it such that $P'Q = QP = QA$, the locus of P and P' is called the *strophoid of S with respect to the pole O and the fixed point A*.

The Right and Oblique Strophoids

These are strophoids of a straight line with respect to a pole not on the line, the fixed point being on the line. For the right strophoid the fixed point is the foot of the perpendicular from the pole to the line. (See ch. 10.)

Freeth's Nephroid

This is the strophoid of a circle with respect to its centre as pole, the fixed point being on the circumference (Fig. 90). The polar equation, with O as pole and OA as initial line, is
$$r = a(1 + 2\sin\tfrac{1}{2}\theta).$$

The curve can be used for describing a regular heptagon as follows: Let the perpendicular bisector of OA meet the outer branch at H and let OH meet the base-circle at Q. Then angle $AOH = \tfrac{3}{7} \times 180°$ and angle $QAO = \tfrac{2}{7} \times 180°$. (*Hint for proof:* Call angle AOH θ and use three isosceles triangles.)

Other Strophoids

The following are suggested for investigation:

1. The strophoid of a straight line with respect to a pole at a distance a from the line, the fixed point being on the perpendicular from the pole to the line (produced beyond the line), at a distance from the line (i) less than a, (ii) equal to a, (iii) greater than a.

2. The strophoid of a circle with respect to a point on the circumference, the fixed point being the diametrically opposite point.

[135]

3. The strophoid of a circle of radius a with respect to its centre, the fixed point being at a distance from the centre (i) less than a, (ii) between a and $2a$, (iii) equal to $2a$, (iv) between $2a$ and $3a$, (v) greater than $3a$.

4. The strophoid of a cardioid with respect to the point opposite the cusp-point as pole, with the fixed point at the cusp.

5. The strophoid of a rectangular hyperbola with respect to the centre as pole, with a vertex as the fixed point.

Freeth's Method for the Regular Nonagon†

Let AB be the base of an equilateral triangle ABC. Let Q be any point on BC produced and let AQ be joined and produced to P, so that $QP = QB$. Then the locus of P is part of an oblique strophoid. If the perpendicular bisector of AB meets the locus of P at P_1, and AP_1 is joined, the angle BAP_1 is 80°. From this an angle of 40° can be found and the regular nonagon constructed. (*Hint for proof:* Call angle BAP_1 θ and use isosceles triangles.)

If the complete strophoid is drawn, the perpendicular bisector of AB meets it at three points, the corresponding values of angle BAP being 80°, 20° and −40°.

Freeth's Supertrisectrix†

This is the strophoid of a trisectrix with respect to the centre of the base-circle as pole and the node as fixed point. Let the trisectrix be drawn as in Figs. 29, 30 (p. 45), with O as centre of the base-circle and A as pole, QP and QP' being equal to QO. OP is now joined and produced both ways. Points R and R' are found on this line such that $PR = PR' = PA$. The locus of R and R' is the supertrisectrix, a closed curve having an outer loop and four inner loops. The four inner loops intersect at O, which is thus a quadruple point. The node of the trisectrix, A, is a quintuple point.

This curve may be used for drawing a regular undecagon. If the perpendicular bisector of AO meets the outer branch at a point R_1, corresponding to P_1 on the trisectrix and Q_1 on the base-circle, the angle AR_1O will be $\frac{1}{11}$ of 180°. For, if this angle $= \theta$,

$$P_1R_1 = P_1A, \qquad \therefore \quad \text{angle } P_1AR_1 = \theta \quad \text{and} \quad \text{angle } OP_1A = 2\theta;$$
$$Q_1P_1 = Q_1O, \qquad \therefore \quad \text{angle } P_1OQ_1 = 2\theta \quad \text{and} \quad \text{angle } OQ_1A = 4\theta;$$
$$OQ_1 = OA, \qquad \therefore \quad \text{angle } OAQ_1 = 4\theta \quad \text{and} \quad \text{angle } OAR_1 = 5\theta;$$
$$R_1A = R_1O, \qquad \therefore \quad \text{angle } AOR_1 = 5\theta.$$

Hence, in triangle OAR_1, the base angles are each five times the vertical angle.

† T. J. Freeth, *Proc. Lond. Math. Soc.* vol. x (1879).

If the perpendicular bisector of AO meets the five branches of the curve at R_1, R_2, R_3, R_4 and R_5, in that order, the vertical angles of the triangles so formed are, respectively, $\frac{1}{11}$, $\frac{3}{11}$, $\frac{5}{11}$, $\frac{7}{11}$ and $\frac{9}{11}$ of $180°$; and the base angles are, respectively, $\frac{5}{11}$, $\frac{4}{11}$, $\frac{3}{11}$, $\frac{2}{11}$ and $\frac{1}{11}$ of $180°$.

Strophoids with the Pole at Infinity

Let S be a given curve and A a fixed point, as before. From any point Q on S a line is drawn parallel to a given fixed direction and points P and P' are marked on it such that $P'Q = QP = QA$. Then the locus of P and P' is the strophoid of Q with respect to the given direction and the fixed point A.

The strophoid (in this sense) of a straight line with respect to the perpendicular direction and a fixed point not on the line is a rectangular hyperbola. That of a circle, centre C, with respect to a fixed point A and the direction CA is a curve consisting, in general, of two ovals. If A is inside the circle, the two ovals touch each other; if A is outside, they are separate, but one of them takes a two-petalled form; and if A is on the circumference, they unite to form a three-petalled curve whose polar equation (with A as pole and AC as initial line) is

$$r = 4a \cos 2\theta \cos \theta.$$

Fig. 91. A seven-cusped epicycloid

17

ROULETTES

Definition

If a curve rolls, without slipping, along another, fixed, curve, any point or line which moves with the rolling curve describes a *roulette*. The locus of a point attached to the rolling curve is a *point-roulette*, and the envelope of a line attached to the rolling curve is a *line-roulette*.

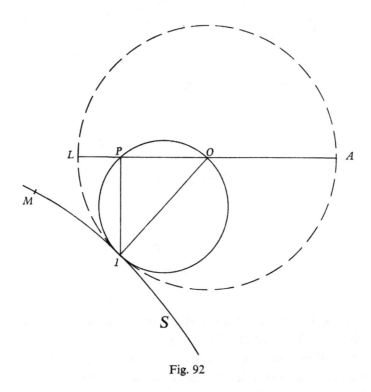

Fig. 92

Thus the cycloid is a point-roulette, since it is the locus of a point on the circumference of a circle which rolls on a fixed straight line; and it has been shown, on p. 86, that it may also be described as a line-roulette. If a circle rolls on the outside or inside of a fixed circle, the roulette traced by a point on its circumference is an epicycloid or hypocycloid. These curves are defined as point-roulettes, but it may be shown that they are also line-roulettes, as follows:

Let P be a point on the circumference of a circle rolling on a fixed curve S (Fig. 92). Let I be the point of contact and IO the diameter through I. The arc IP is equal to an arc IA of the curve S, where A is a fixed point. Now consider the circle whose centre is O and whose radius is OI. Let PO be produced to meet this circle at L and M. Then arc $IL =$ arc IP (radius double, angle at centre half).

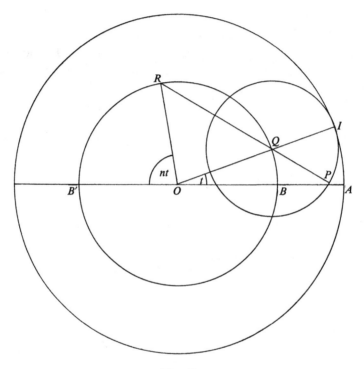

Fig. 93

Therefore, if this new circle were rolled along S, L would eventually arrive at A It follows that LM is a diameter fixed relative to the rolling circle. Moreover, as this circle rolls, its instantaneous centre is at I and, since angle IPO is a right angle, P is the point of contact of LM with its envelope. The locus of P is, therefore, the same as the envelope of LM.

Thus the point-roulette of a point on the circumference of a circle rolling along any curve is the same as the line-roulette of the diameter of a circle of twice the radius.

To Draw any Hypocycloid

To draw a hypocycloid having $(n+1)$ cusps, draw a circle, centre O, and a diameter $B'OB$ (Fig. 93). Draw radii OQ, OR, such that angle $BOQ = t$ and angle $B'OR = nt$, these angles being measured in opposite senses. Join RQ. Then, as t varies, the envelope of RQ will be the required hypocycloid.

Proof: If the radius of the original circle is $(n-1)a$, draw another circle with

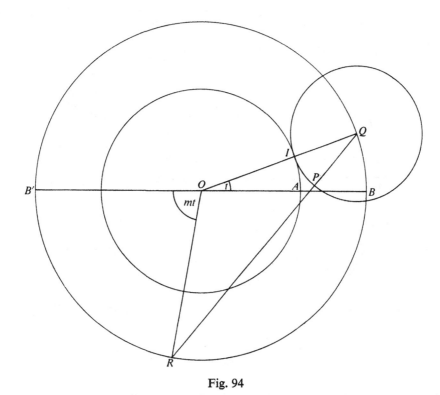

Fig. 94

centre O and radius $(n+1)a$, cutting OB produced at A. With centre Q draw a circle of radius $2a$, touching the second circle at I and meeting RQ produced at P. Then arc IP = arc IA: for IP subtends an angle $\frac{1}{2}(n+1)t$ at the centre of a circle of radius $2a$ and IA subtends an angle t at the centre of a circle of radius $(n+1)a$. Therefore, if the circle whose centre is Q rolls round the inside of the circle whose centre is O and whose radius is OA, P will always arrive at A. Thus P is a fixed point of the rolling circle and the diameter through P is a line carried by the rolling circle. The envelope of RQ is, therefore, that of a diameter of the rolling circle and

[141]

is the same as the locus of a point on the circumference of a circle of radius half as great, i.e. of radius a, rolling on the same curve, namely the circle of radius $(n+1)a$. This locus is a hypocycloid having $(n+1)$ cusps.

To Draw Any Epicycloid

To draw an epicycloid having $(m-1)$ cusps, follow the same method with n changed to $-m$ (i.e. angle $B'OR$ should be of magnitude mt, measured this time in the same

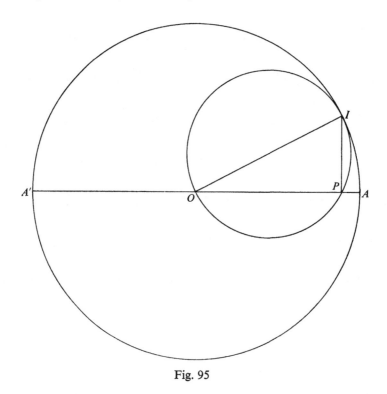

Fig. 95

sense as angle AOI) (Fig. 94). The proof follows similar lines, the radius of the original circle being now $(m+1)a$.

Various Values of n and m

When $n = 1$, the above construction fails. But, if a circle rolls on the inside of a circle of radius twice as great, the locus of a point on its circumference is a straight line. In Fig. 95, IP is drawn perpendicular to a diameter $A'A$ of a fixed circle. P then lies on the circle on OI as diameter, and arc IP = arc IA. Thus the two-cusped hypocycloid is a segment of a straight line.

If the value of n is a rational fraction, $(n+1)$ is still the ratio of the circumference of the fixed circle to that of the rolling circle; but, if $(n+1) = p/q$, where p and q are integers prime to each other, there will be p cusps, produced in q revolutions of the radius OI. The same is true for epicycloids if $(m-1) = p/q$. In Fig. 91, $m = \frac{10}{3}$.

Exercises

1. What curves correspond to the values $n = 2$, $n = 3$, $m = 2$, $m = 3$?
2. Draw the curves corresponding to some other values, such as $n = 4$, $n = \frac{3}{2}$, $m = 4$, $m = \frac{9}{2}$.

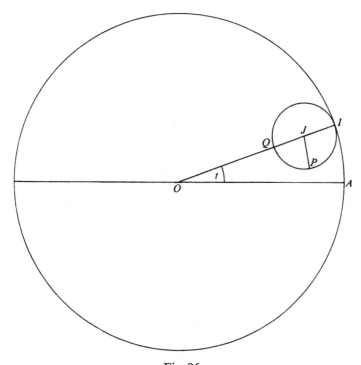

Fig. 96

Parametric Equations

In Fig. 96, J is the centre of a circle of radius a rolling on the inside of a fixed circle whose centre is O and whose radius is $(n+1)a$. P is a point tracing a hypocycloid of $(n+1)$ cusps, starting at A. Then, if angle $AOI = t$, arc PI = arc AI, and angle $PJI = (n+1)t$. Hence the inclination of JP to OA is $-nt$. With O as origin and OA as axis of x, parametric equations for the hypocycloid are

$$x = na\cos t + a\cos nt, \quad y = na\sin t - a\sin nt.$$

Corresponding equations for the epicycloid are

$$x = ma\cos t - a\cos mt, \quad y = ma\sin t - a\sin mt.$$

[143]

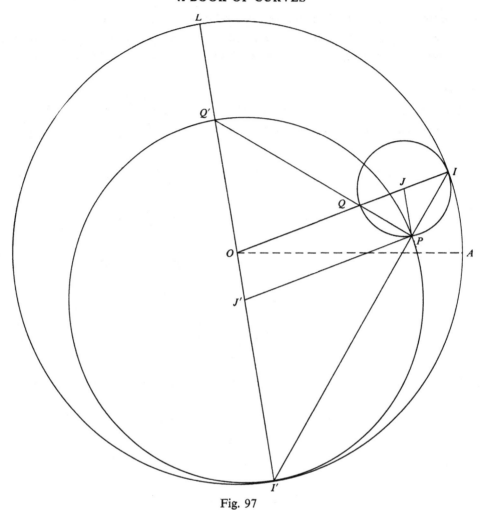

Fig. 97

Double Generation

* Every hypocycloid and every epicycloid can be generated as a locus in two ways. Consider, in Fig. 97, the hypocycloid traced by the point P on the circle whose centre is J and whose radius is a, as that circle rolls on the inside of the fixed circle whose centre is O and whose radius is $(n+1)a$. Let IP produced meet the fixed circle at I'; and let $I'O$ meet PQ produced at Q', and the fixed circle again at L. Then the isosceles triangles IOI' and IJP are similar; therefore JP is parallel to OI'. It follows that triangles QOQ' and QJP are similar, and hence that $OQ' = OQ$. (Q' is in fact the point previously called R.) $\therefore\ I'Q' = (n+1)a+(n-1)a = 2na$,

a fixed length. A circle drawn on $I'Q'$ as diameter passes through P, its centre being J', where $OJ' = a = JP$; hence $JPJ'O$ is a parallelogram. The arcs $I'P$, PI and $I'I$ all subtend equal angles at the centres of their respective circles and, as the radii of the first two circles are together equal to that of the third, it follows that arc $I'P + $ arc $PI = $ arc $I'I$.

Now P is a point fixed on the circumference of the circle whose centre is J. If it eventually reaches the fixed circle at A, the point of the other circle which was at P must also arrive at A. Therefore, P is a point fixed on the circumferences of both the rolling circles and the hypocycloid is traced out in two ways.

It may be noted that $I'A : AI = n : 1$; and that, for the hypocycloid to be traced simultaneously by the two methods, the angular velocities of the circles whose centres are J and J' respectively must be in that same ratio. (*Proof*: If angles $I'OA$ and AOI are nt and t respectively, JP must turn into the position OA, a turn of amount nt; and $J'P$ must also turn into the position OA, a turn of amount t.)

Evolute of the Hypocycloid

* Since I is the instantaneous centre of one of the rolling circles, $I'I$ is a normal to the curve, and the evolute is the envelope of $I'I$. Let DD' be a diameter of the fixed circle such that angle $AOD = 1/(n+1) \times 180°$ and angle $AOD' = n/(n+1) \times 180°$ (D and I being on one side of A, with D' and I' on the other). Then

$$\text{angle} \quad I'OD' = n \times \text{angle } IOD;$$

and it follows that the envelope of $I'I$ is a hypocycloid having $(n+1)$ cusps (if n is an integer), DD' being one of the cusp lines. The evolute is thus a curve similar to the original hypocycloid, enlarged in the ratio $OI : OQ$, i.e. $(n+1) : (n-1)$. (See Fig. 98.)

Length of Arc

* Let a be the radius of the rolling circle and $(n+1)a$ that of the fixed circle. Then, if A is a cusp of the original curve and A' a cusp of the evolute, and if OA' meets the original curve at M (Fig. 99),

$$OA = (n+1)a, \quad OA' = \frac{n+1}{n-1}OA = \frac{(n+1)^2}{n-1}a \quad \text{and} \quad OM = (n-1)a.$$

Therefore $\quad MA' = OA' - OM = \dfrac{(n+1)^2}{n-1}a - (n-1)a = \dfrac{4na}{n-1}.$

But the arc $A'A$ of the evolute is equal to $A'M$. Hence the whole length of the evolute (if n is an integer) is

$$2(n+1)\frac{4na}{n-1}, \quad \text{i.e.} \quad 8na\frac{n-1}{n+1}.$$

The whole length of the original curve is therefore $8na$ (or $8qna$ if $n = p/q$).

For an epicycloid, where a is the radius of the rolling circle and $(m-1)a$ that of the fixed circle, the whole length of the curve is $8ma$ (or $8qma$, if $m = p/q$).

Area

* It is supposed that n and m are integers. In Fig. 97, P is a point on the hypocycloid and PQQ' is the tangent at P, cutting the inscribed circle at Q and Q'. If PI, the normal, touches the evolute at E,

$$\frac{EI}{II'} = \frac{PQ}{QQ'} = \frac{PJ}{OQ'} = \frac{1}{n-1}.$$

Also,

$$\frac{IP}{II'} = \frac{JP}{OI'} = \frac{1}{n+1}.$$

Therefore,

$$\frac{EI}{IP} = \frac{n+1}{n-1} \quad \text{and} \quad \frac{EI}{EP} = \frac{n+1}{2n}.$$

By the argument used on p. 60, the area between the evolute and the circle $I'AI$ is a fraction $(n+1)^2/(4n^2)$ of the area between the two hypocycloids. If the area of the original hypocycloid is S, that of the evolute is $S(n+1)^2/(n-1)^2$, and that of the space between is $4nS/(n-1)^2$. Therefore,

$$\frac{(n+1)^2}{(n-1)^2}S - \pi(n+1)^2 a^2 = \frac{(n+1)^2}{4n^2} \cdot \frac{4n}{(n-1)^2} S,$$

whence $S = \pi a^2(n^2 - n)$.

By a similar method, or by changing n into $-m$, the area of the epicycloid is $\pi a^2(m^2 + m)$.

The Prolate and Curtate Cycloids

If a circle of radius a rolls along a straight line, the roulette traced by a point carried by it, distant h from its centre, is a *prolate cycloid* if $h > a$, and a *curtate cycloid* if $h < a$. Parametric equations for these curves are:

$$x = at - h\sin t, \quad y = a - h\cos t.$$

(Cf. the cycloid, p. 82.)

These curves may be drawn by suitable modifications of the methods used for the cycloid (e.g. in the *Second Method*, all radii should be h, while the intervals between the points remain as $0.35a$). The prolate cycloid has a node and loop for every revolution of the rolling circle.

Epitrochoids and Hypotrochoids

The term *trochoid* has the same meaning as *roulette*. It is used more particularly for the roulettes traced by points carried by a circle rolling on a fixed circle. These

Fig. 98

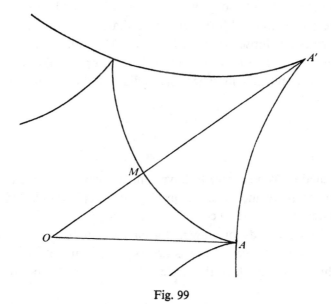

Fig. 99

[147]

are called *epitrochoids* or *hypotrochoids* according as the circle rolls on the outside or inside of the fixed circle.

Such curves can be drawn by extensions of the methods described on pp. 141, 142 for hypocycloids and epicycloids. In Fig. 97 it is necessary to find a point T on JP (produced if necessary) such that $JT = h$. Hence the following method:

Draw a circle of radius na and a diameter $U_0 O J_0$. From J_0 mark off points $J_1, J_2, ...,$ on the circumference at 5° intervals. From U_0 mark $U_1, U_2, ...,$ at intervals of $5n°$, in the opposite sense to $J_1, J_2, ...,$ for a hypotrochoid, in the same sense as $J_1, J_2, ...,$ for an epitrochoid. Draw radii $OU_1, OU_2, ...$ From $J_0, J_1, J_2, ...,$ draw lines $J_0 T_0, J_1 T_1, J_2 T_2, ...,$ of fixed length h, parallel to $U_0 O, U_1 O, U_2 O, ...,$ for a hypotrochoid, or to $OU_0, OU_1, OU_2, ...,$ for an epitrochoid. Then (if n is an integer), $T_0, T_1, T_2, ...,$ will lie on a hypotrochoid, having $(n+1)$ loops if $h > a$; or an epitrochoid, having $(n-1)$ loops if $h > a$.

Parametric equations for the hypotrochoid are:

$$x = na\cos t + h\cos nt, \quad y = na\sin t - h\sin nt;$$

and for the epitrochoid:

$$x = ma\cos t - h\cos mt, \quad y = ma\sin t - h\sin mt.$$

It was shown (p. 48) that the limaçon is an epitrochoid. The ellipse may be regarded as a hypotrochoid, formed when the diameter of the rolling circle is half that of the fixed circle. (*Hint for proof:* If the tracing-point P lies on a diameter QQ' of the rolling circle, Q and Q' will describe two-cusped hypocycloids, i.e. diameters of the fixed circle. Moreover, these diameters will be at right angles. Hence P describes an ellipse, as in the Trammel Method, p. 19.)

The involute of a circle may be regarded as an epicycloid, and the Spiral of Archimedes as an epitrochoid, when the radius of the rolling circle is infinite. (Fig. 113, p. 173.)

Other Roulettes

To draw a roulette it is necessary to have some measure of the relative lengths of arcs on the rolling curve and the fixed curve. For the epicycloids and hypocycloids these lengths were compared by considering the angles subtended at the centres of the circles; and the methods for drawing the curves were based on the relationships of those angles. For the cycloid it was necessary to mark off some multiple of π on the straight line along which the circle rolls. Not many other roulettes can be easily drawn.

It has been mentioned (p. 123) that the point-roulette of the focus of a parabola, when the parabola rolls along a straight line, is a catenary. This can be proved by the methods of the calculus, but a drawing cannot easily be made because there is no convenient measure of the length of arc of the parabola.

It has been shown (p. 103) that the roulette traced by the pole of an equiangular spiral, when the spiral rolls along a straight line, is another straight line.

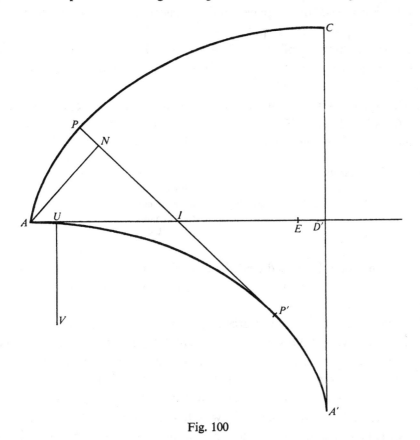

Fig. 100

Drawing Exercises

1. If an arch of a cycloid is rolled along a straight line, the roulette of the mid-point of the arch can be drawn. For suppose that the arc AA' (Fig. 100) is half of a cycloidal arch and is the evolute of AC, another such arc. If PP' is a tangent to the one and a normal to the other, then $PP' = $ arc AP'; and the effect of rolling the arc AA' along the straight line AD' is to bring PP' into the position AE. If AN is drawn perpendicular to PP', and if AU, UV are drawn along and per-pendicular to AD', equal respectively to PN, NA, then V will be a point of the roulette.

[149]

Suitable Dimensions

With the paper in the 'landscape' position, place A rather below the centre of the paper and draw two arches of the cycloid AC to the right of A. A large number of positions of the line PI must be shown. The positions of V may be quickly plotted by placing a piece of paper with two of its edges along NP and NA, the lengths PN and NA being marked on the paper and so transferred to the positions AU and UV. The locus is a spiral, with its pole at A.

If an existing drawing of one arch of a cycloid is used, the first half-turn of the spiral can be drawn, provided there is space below A' equal to about three-quarters of $A'D'$.

2. Draw a regular octagon $ABCDEFGH$, inscribed in a circle of diameter d. Join AC, AD, AE, AF, AG. Draw the roulette of the point A when the octagon rolls along a fixed straight line.

Suitable Dimensions

		Diameter d	
Paper:	1_L	2 in. or	6 cm.
	2_L	3 in.	8 cm.
	3_L	4 in.	10 cm.

The octagon should be placed near the top left-hand corner, orientated so that AB is its base. The fixed line for the roulette should be near the bottom edge of the paper.

The roulette consists of seven arcs of circles (for one revolution of the octagon) and may be drawn as follows: Mark points a_1, b, c, d, e, f, g, h, a_8 on the fixed line at intervals equal to AB. With centre b and radius ba_1, draw an arc a_1a_2 of 45°; with centre c and radius ca_2, draw an arc a_2a_3 of 45°; and so on.

It may be proved that the area between the seven arcs and the base-line is equal to that of the octagon together with twice that of its circumscribing circle. (*Hint*: The area is divided into six triangles and seven sectors of circles. The six triangles are together equal to the area of the octagon. To prove that the seven sectors are together equal to twice the area of the circle, note that

$$AB^2 + AF^2 = AC^2 + AG^2 = AD^2 + AH^2 = d^2.)$$

A similar result holds good from a polygon of $2n$ sides; and from this the area under a cycloidal arch may be deduced.† For, if n is increased, the polygon approximates more and more closely to a circle, and the roulette to a cycloid. Hence the area under a cycloidal arch is three times that of the rolling circle.

† W. Hope-Jones, *Mathematical Gazette*, vol. x, p. 207.

Hypocycloids and Epicycloids: Summary

1. *For a hypocycloid:* Let a be the radius of the rolling circle and $(n+1)a$ that of the fixed circle. With the centre of the fixed circle as origin and one cusp on the x-axis, parametric equations are:

$$x = na\cos t + a\cos nt, \quad y = na\sin t - a\sin nt.$$

2. *For an epicycloid:* Let a be the radius of the rolling circle and $(m-1)a$ that of the fixed circle. With the centre of the fixed circle as origin and one cusp on the x-axis, parametric equations are:

$$x = ma\cos t - a\cos mt, \quad y = ma\sin t - a\sin mt.$$

3. The number of cusps is $(n+1)$ or $(m-1)$, if these numbers are integers; but if one of them is a fraction p/q, where p and q are integers prime to each other, there are p cusps obtained in q revolutions of the line of centres.

Some special cases:

    $n = 1$. Segment of a straight line.

    $n = 2$. Deltoid. $m = 2$. Cardioid.

    $n = 3$. Astroid. $m = 3$. Nephroid.

4. $L = 8na$ or $8ma$, if n, m are integers; but $8qna$ or $8qma$, if they are rational fractions.

5. $A = \pi a^2(n^2 - n)$ or $\pi a^2(m^2 + m)$, where n, m are integers.

6. The evolute is a similar curve, larger in the linear ratio $(n+1):(n-1)$, or smaller in the linear ratio $(m-1):(m+1)$.

7. For the hypocycloid, the radius of the rolling circle may be either a or na, that of the fixed circle being $(n+1)a$; and for the epicycloid the radius of the rolling circle may be a or ma, that of the fixed circle being $(m-1)a$.

8. The hypocycloid is the envelope of a diameter of a circle of radius $2a$ rolling on the inside of a fixed circle of radius $(n+1)a$; and the epicycloid is the envelope of a diameter of a circle of radius $2a$ rolling on the outside of a fixed circle of radius $(m-1)a$.

Roulette ('shaped like a wheel') was Roberval's name for the cycloid, before *cycloid* became universally accepted. *Trochoid* is the Greek word for 'wheel-shaped'.

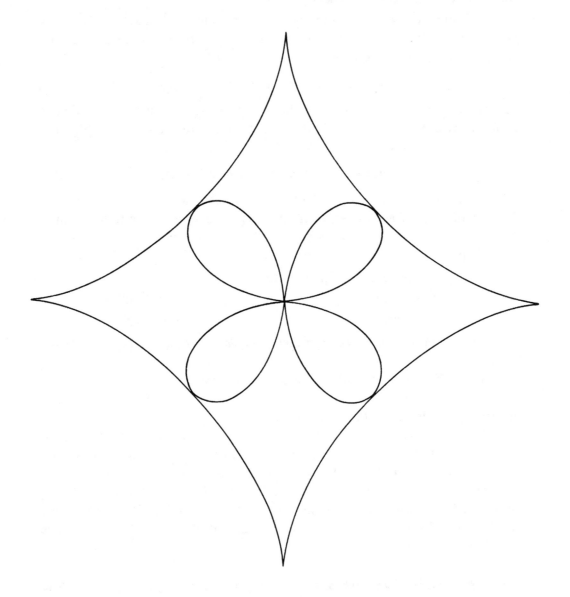

Fig. 101. The astroid and the quadrifolium

· 18

PEDAL CURVES

Definition of a Pedal Curve

If S is any curve and O is a fixed point (called the *pedal-point*), the locus of the foot of the perpendicular from O to a variable tangent to the curve is called the *pedal of S with respect to O*.

The cardioid and limaçon are pedals of a circle with respect to a point which does or does not lie on the circumference. The pedal of an ellipse with respect to one of its foci is the auxiliary circle (p. 13); the pedal of a rectangular hyperbola with respect to its centre is a lemniscate (p. 112).

To Draw a Pedal Curve

If the original curve has been drawn as an envelope the pedal can at once be plotted as a locus. (A set square may conveniently be used.) A curve geometrically similar to the pedal curve may be drawn as an envelope, as follows: With centre at any point Q of S, and radius QO, draw a circle: the envelope of such circles will be a curve similar to the pedal curve, on double scale. This curve is called the *orthotomic of S with respect to O*. The method was used for drawing the cardioid (p. 35), the limaçon (p. 49) and the lemniscate (p. 112).

It may be noted that, with the pedal-point as origin, the angle ϕ between the radius vector and the tangent is the same for corresponding points of the pedal and the original curve. (See p. 14.)

Rose-Curves

These are the pedals of the epicycloids and hypocycloids with respect to their centres. If the original curve is a hypocycloid, there is one 'leaf' of the pedal opposite each section (i.e. between each pair of adjacent cusps) of the original.

If the original curve is an epicycloid, the pedal is of the same nature, though different in appearance. There is again one leaf opposite each section of the original curve, but the leaf is 180° wider, in angular measure, than the corresponding section of the original. The result is that adjacent leaves overlap each other. In the case of a nephroid, each leaf is similar in appearance to a cardioid; and if the original curve is a cardioid, the pedal consists of a single leaf so wide as to overlap itself.

[153]

Other Pedal Curves

Pedals of the following curves are suggested for drawing:

    1. The parabola with respect to its vertex (*the Cissoid of Diocles*).

    2. The parabola with respect to the image of the focus in the directrix (*the Trisectrix of Maclaurin: see below*).

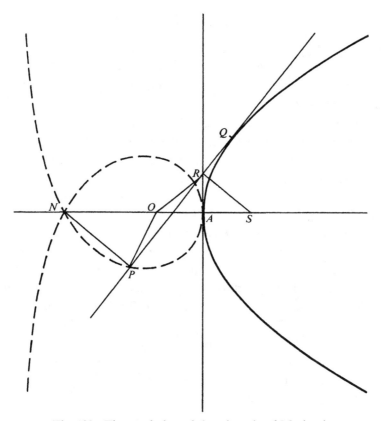

Fig. 102. The parabola and the trisectrix of Maclaurin

    3. The (ordinary) trisectrix (p. 46) with respect to the centre of its base-circle.

    4. The (ordinary) trisectrix with respect to its node.

    5. The cardioid with respect to its cusp-point (*Cayley's Sextic*).

    6. Epicycloids and hypocycloids with respect to their centres.

    7. The deltoid with respect to various points on the curve or inside it.

The Trisectrix of Maclaurin

* Let the focus and vertex of the parabola be S and A respectively, and let O and N be points on the axis such that $NO = 2a$, $OA = AS = a$ (Fig. 102). Let Q be any

[154]

point on the parabola such that the perpendicular NP from N to the tangent at Q gives a point P on the loop of the pedal curve. Then angle $ANP = \frac{1}{3} \times$ angle AOP. (*Hint for proof:* If the tangent at the vertex A is cut by PQ at R, SR is perpendicular to PQ (p. 3). Moreover, O is equidistant from PN and SR; hence

$$OP = OR = RS.$$

Call angle ANP x.)

With O as pole and OA as initial line, the equation of the pedal curve (the *Trisectrix of Maclaurin*) is $r = a \sec \frac{1}{3}\theta$. At N the curve makes angles of $\pm 60°$ with NO.

** The area of the loop is $3\sqrt{3}a^2$. The inverse with respect to the node is a hyperbola. The Cartesian equation, with N as origin and NA as axis of x, is

$$(x+a)y^2 = x^2(3a-x).$$

Some Curves and their Pedals

Curve	Pedal-point	Pedal
Circle	Point on circumference	Cardioid
Circle	Any other point	Limaçon
Parabola	Focus	Straight line
Parabola	Vertex	Cissoid
Parabola	Foot of directrix	Right strophoid
Parabola	Other point of directrix	Oblique strophoid
Parabola	Image of focus in directrix	Trisectrix of Maclaurin
Ellipse or hyperbola	Focus	Auxiliary circle
Rectangular hyperbola	Centre	Lemniscate
Epicycloids and hypocycloids	Centre	Rose-curves
Cardioid $(r^3 = 4ap^2)$	Cusp-point	Cayley's Sextic $(r^4 = 4ap^3$, or $r = 4a\cos^3 \frac{1}{3}\theta)$
Deltoid	Cusp	Simple folium or ovoid $(r = 4a\cos^3 \theta)$
Deltoid	Vertex	Double folium $(r = 4a\cos\theta\sin^2\theta)$
Deltoid	Any other point on the curve	Double folium (unsymmetrical)
Deltoid	Any other point on the inscribed equilateral triangle	Trifolium
Cissoid	Point on axis, beyond asymptote, whose distance from cusp is four times that of asymptote	Cardioid
Equiangular spiral	Pole	An equal spiral
Sinusoidal spiral $(r^{n+1} = a^n p)$	Pole	Another sinusoidal spiral $(r^{2n+1} = a^n p^{n+1})$
Involute of circle	Centre of circle	Archimedian spiral

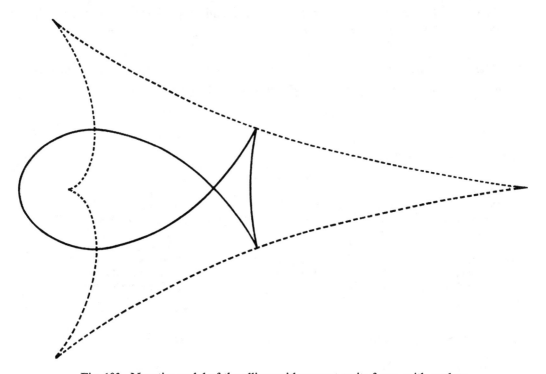

Fig. 103. Negative pedal of the ellipse with respect to its focus, with evolute

19

NEGATIVE PEDALS

Definition

Let S be any curve and let O be a fixed point. If Q is a variable point on the curve and a line is drawn through Q at right angles to OQ, the envelope of this line is called the *negative pedal of S with respect to O*.

The parabola is the negative pedal of a straight line (p. 3); the ellipse and hyperbola are negative pedals of a circle with respect to a point inside or outside it (pp. 13, 25).

Other Negative Pedals

1. The negative pedal of a cardioid with respect to the point on the curve directly opposite the cusp-point is the Cissoid of Diocles. (The distance of the cusp-point of the cissoid from O is $\frac{4}{3}$ as great as that of the cusp-point of the cardioid.)

2. The negative pedal of the parabola with respect to its focus is *Tschirnhausen's Cubic*, also known as *l'Hospital's Cubic*, or the *Trisectrix of Catalan*. (See below.)

3. The negative pedal of the ellipse with respect to its focus is an egg-shaped curve if $e \leqslant \frac{1}{2}$; but, if $e > \frac{1}{2}$, it has a node and two cusps, as shown in Fig. 103. In the special case $e^2 = \frac{1}{2}$, illustrated in Fig. 103, the node is rectangular and the distance between the cusps is equal to the maximum width of the other part of the curve, both these distances being equal to half the major axis of the ellipse. (For further details see the *Mathematical Gazette*, vol. XLI, p. 254.)

4. The negative pedal of the ellipse with respect to its centre is a curve having two nodes and four cusps, provided $e^2 \geqslant \frac{1}{2}$. It is known as *Talbot's Curve*.

5. The negative pedal of a hyperbola with respect to its centre is a curve having two nodes and two asymptotes at right angles to those of the hyperbola.

Suitable Dimensions	Radius of base-circle	Distance of focus from centre
Paper: 1_P	1 in. or 3 cm.	2 in. or 6 cm.
2_P	1·5 in. 4 cm.	3 in. 8 cm.
3_P	2 in. 5 cm.	4 in. 10 cm.

The focus should be on a level with the centre.

6. The negative pedal of a nephroid with respect to one of its cusp-points is a curve having two cusps and one asymptote, shaped like a Greek letter Ω.

Tschirnhausen's Cubic

Let the focus and vertex of the parabola be S and A respectively. If Q is a point on the parabola, and P the corresponding point of the cubic (Fig. 104),

$$\text{angle } ASQ = \tfrac{2}{3} \times \text{angle } ASP.$$

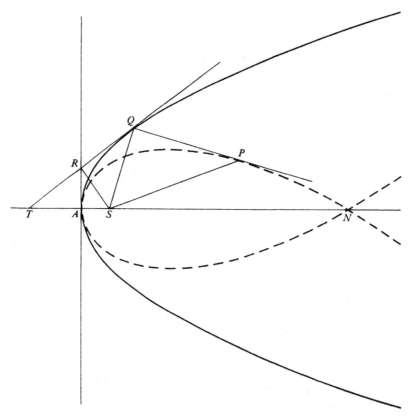

Fig. 104. The parabola and Tschirnhausen's cubic

(*Hint for proof:* Let the tangent to the parabola at Q meet the axis at T. Then angle SPQ = angle SQT = angle STQ.)

With S as pole, and SA as initial line (so that angle $ASP = \theta$), the polar equation of the cubic is $r = a\sec^3(\theta/3)$. (*Hint for proof:* Use the same figure, with SA equal to a.)

The pedal equation, with S as pole, is $ar^2 = p^3$.

[158]

Some Curves and their Negative Pedals

Curve	Point	Negative pedal
Straight line	Point not on the line	Parabola
Circle	Point inside	Ellipse
Circle	Point outside	Hyperbola
Circle, radius a	Point distance $a\sqrt{2}$ from centre	Rectangular hyperbola
Parabola	Focus	Tschirnhausen's cubic
Ellipse	Centre	Talbot's Curve
Ellipse	Focus	Curve shown in Fig. 103
Cardioid	Cusp	Circle
Cardioid	Point opposite cusp	Cissoid
Limaçon	Node or pole	Circle
Equiangular spiral	Pole	Equiangular spiral
Sinusoidal spiral	Pole	Sinusoidal spiral
Archimedian spiral	Pole	Involute of circle

The names *Tschirnhausen's Cubic* and *Cayley's Sextic* (p. 155) are due to R. C. Archibald's attempt to classify the curves in a paper published at Strasbourg in 1900.

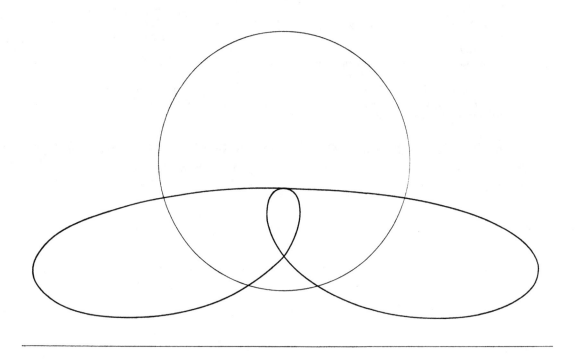

Fig. 105. Glissette of the mid-point of a rod whose ends slide
on a straight line and a circle

20

GLISSETTES

Definition

When a curve, supposed rigid, slides against two fixed curves (i.e. when it is moved so that it always touches the fixed curves), the locus of any point, or the envelope of any line or curve, attached to the sliding curve is called a *glissette*.

If an ellipse slides against two fixed perpendicular lines, the locus of its centre is

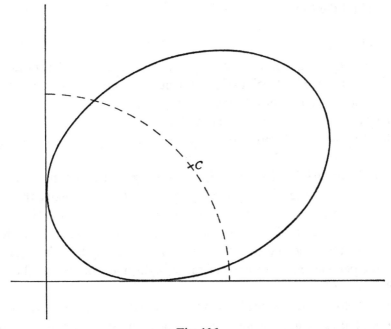

Fig. 106

an arc of a circle (Fig. 106). (For it is a well-known property of the ellipse, which may be proved or verified by drawing, that two tangents at right angles intersect at a fixed distance from the centre.)

If a segment of a straight line slides against two fixed perpendicular lines, the locus of its mid-point is a circle. This is easily proved independently, but it may also be regarded as a special case of the preceding result, the segment being regarded as the limiting case of a very narrow ellipse. The locus of any other fixed point of the segment is an ellipse (p. 19, Ex. 3, with P' on PN produced); and the envelope of the segment is an astroid (p. 53).

If a set square slides with two of its sides each passing through a fixed point, the locus of the vertex in which those sides meet is a circle. (Here the sliding curve consists of a pair of straight lines, and the fixed curves are reduced to points.) If the sides each touch one of two fixed circles, the locus consists of four limaçons or two cardioids. (See p. 50.)

The right strophoid was drawn first as a glissette (p. 91), with a line and a point sliding against a point and a line respectively.

Any conchoid may be regarded as a glissette, with a line and one of its points sliding respectively against the given fixed point and the given curve.

Another special kind of glissette is produced when a curve moves so as to touch a fixed curve at a fixed point. Here the 'two fixed curves' of the definition are the fixed curve and the fixed point on it.

Other Examples

The following are suggested for drawing or investigation by other means:

1. The locus of the mid-point of a segment of a straight line whose ends move on two intersecting lines not at right angles.

2. The envelope of the same segment of a line.

3. The locus of the mid-point of a segment of a straight line whose ends move on two circles of equal radius (*Watt's Curve*).

Note: If the centres are A, B, the length of the segment should be (i) considerably less than AB, (ii) just less than AB, (iii) equal to AB, (iv) greater than AB. The method given on p. 114 for drawing the lemniscate was a particular case of (iii).

4. The locus of the mid-point of a segment of a straight line of which one end moves on a straight line and the other on a circle, the length of the segment being equal to the perpendicular distance of the straight line from the furthest point of the circle (Fig. 105).

5. The envelope of the same segment of a line. This curve has two cusps and two asymptotes, the curve approaching one end of each asymptote on both sides.

6. The locus of the mid-point of a segment of a straight line of which one end moves on a parabola and the other on its directrix.

7. The envelope of a side of a triangle when the other two sides pass through fixed points or touch fixed circles. This envelope is a circle whose centre may be found as follows: Let the sides AB, AC of the triangle touch circles whose centres are O, Q respectively (Fig. 107). Draw OA' and QA' parallel to BA and CA, to meet at A'. Draw a line through A' parallel to BC to meet circle OQA' at K. Then

K is a fixed point and the perpendicular distance from K to BC is constant. (*Hints for proof:* The circle OQA' is fixed and angle $OA'K$ is constant. The distance from K to BC is the same as that of A'; but A' is fixed relative to BC.)

Another method of proof is to locate the instantaneous centre, I, of the moving triangle and to note that KI is at right angles to BC. The point where BC touches the envelope is the intersection of KI and BC. BC always moves at right angles to KI.

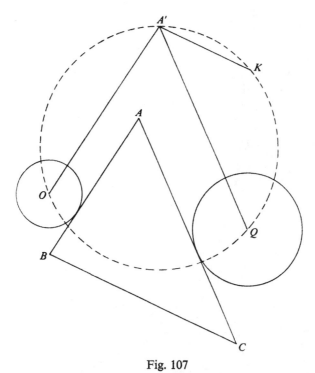

Fig. 107

8. The locus of the focus of a parabola which slides against two perpendicular fixed lines. (*Hint:* Tangents to a parabola which are at right angles meet on the directrix. The polar equation of this glissette, taking the point where the fixed lines meet as origin, and one of them as initial line, is $r = 2a\operatorname{cosec}2\theta$.)

9. The locus of the focus of a parabola which touches a fixed line at a fixed point. The polar equation, taking the fixed point as origin and the fixed line as initial line, is $r = a\operatorname{cosec}^2\theta$.

** 10. *Dürer's 'Conchoid'*. In Dürer's *Underweysung der Messung*, published at Nuremburg in 1525, there is described a curve which he calls *ein muschellini*. It is not, however, a conchoid in the sense defined above.

In Fig. 108, the numbers $1, 2, 3, ..., 16$, are equally spaced, as units, along two axes

[163]

at right angles to each other. The points with corresponding numbers are joined by straight lines, each of which is produced to a total length of sixteen units. The curve so formed is a small part of a quartic curve having two parallel asymptotes and two finite nodes (or, we may say, three nodes, of which one is at infinity).

The complete curve may be drawn as follows: Let Q and R be points (q,o) and (o,r) on the x- and y-axes respectively, such that $q+r = 13$. On QR, produced in

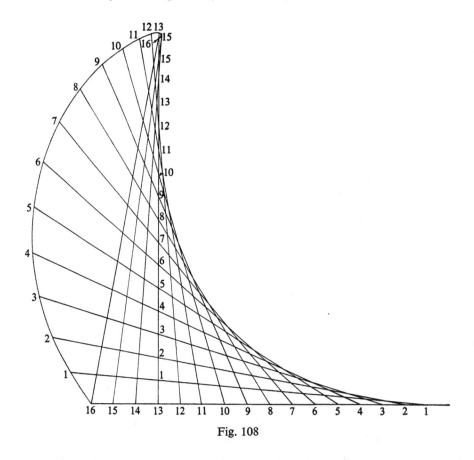

Fig. 108

both directions, mark points P and P' whose distances from Q are sixteen units. The locus of P and P' is the curve.

Suitable dimensions. It is desirable to show points on the x-axis from -25 to $+40$, on the y-axis from -25 to $+25$. For paper 1_L the unit may be $\frac{1}{8}$ in. or 3 mm.; for papers 2_L and 3_L it may be $\frac{1}{5}$ in. or 5 mm.

In Fig. 108 it will be seen that the envelope of the straight lines is a parabola (see p. 7). The 'conchoid' is therefore a point-glissette, formed by a line and one of its points sliding respectively against a parabola and one of its tangents.

[164]

The equation of the quartic curve of which the 'conchoid' is a part may be found by eliminating q and r from

$$x/q + y/r = 1, \quad y^2 + (q-x)^2 = a^2 \quad \text{and} \quad q+r = b,$$

where $a = 16$ and $b = 13$ for the particular case described by Dürer. The result of the elimination is

$$2y^2(x^2+y^2) - 2by^2(x+y) + (b^2-3a^2)y^2 - a^2x^2 + 2a^2b(x+y) + a^2(a^2-b^2) = 0.$$

The asymptotes are $y = \pm a/\sqrt{2}$.

Three special cases are of interest:

(i) If $a = 0$, the curve reduces to a pair of coincident straight lines, $y^2 = 0$;

(ii) if $b = 0$, it reduces to a pair of parallel lines, $y = \pm a/\sqrt{2}$, together with a circle, $x^2 + y^2 = a^2$;

(iii) if $a = b$, one of the nodes becomes a cusp, at the point $(0, b)$.

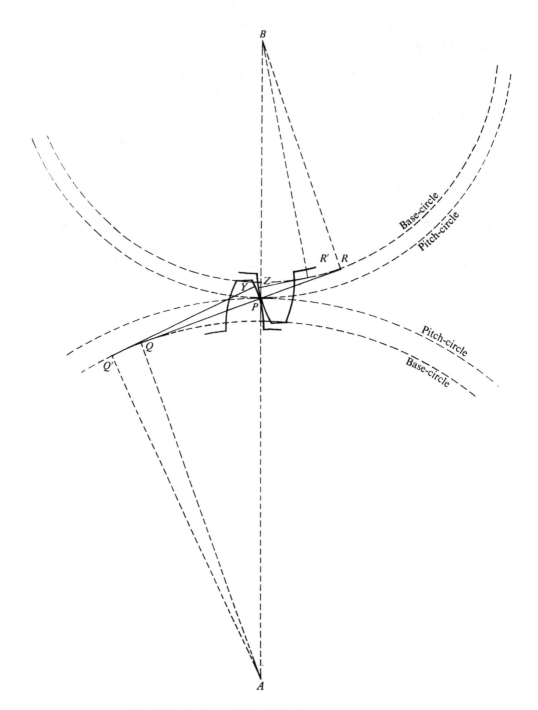

Fig. 109. Involutes of circles in the design of gear teeth

21

EVOLUTES AND INVOLUTES

Definitions

If the normals at points Q and q of a curve meet at C, then the limiting position of C, as q approaches Q, is called the *centre of curvature of the curve at* Q.

The locus of the centre of curvature, as Q varies on the curve, is called the *evolute* of the curve. The original curve is called an *involute* of the new one.

The evolute may alternatively be defined as the envelope of the normal to the curve; for C lies on two tangents to this envelope and, as they approach coincidence, the limiting position of C is a point on the envelope.

Examples of Evolutes

Many examples have already been given, such as the evolute of the parabola (Fig. 1) and that of the cycloid (Fig. 57). It has been shown (p. 105) that the evolute of an equiangular spiral is an equal spiral; and the evolute of any hypocycloid (p. 145) or epicycloid is a curve similar to the original.

Drawing of Evolutes

In the drawing of evolutes it is a help to know that the evolute passes through any ordinary cusp-point of the original curve; that points of inflexion on the original curve correspond to points at infinity on the evolute; and that points of maximum or minimum curvature correspond to cusps of the evolute. (See Fig. 103.)

To draw an evolute it is necessary to have some means of drawing accurately a number of normals to the original curve. Sometimes this can be done by a knowledge of the geometry of the curve as, for example, for the parabola (p. 7), and the cycloidal curves (p. 145).

The evolutes of roulettes and glissettes can usually be drawn, because the position of the instantaneous centre is known and the normal can therefore be drawn. This applies to such curves as the right strophoid and all conchoids and negative pedals.

The following curves are suggested among those whose evolutes can be drawn:

1. *The ellipse*. With any point on the minor axis as centre draw a circle passing through the foci, cutting the curve at P and the further part of the minor axis at G. Then PG is a normal to the ellipse at P.

2. *The limaçon.* With the notation of Fig. 29 (p. 45) the instantaneous centre of *PP'*, regarded as a moving rod, is on the base-circle, at the point *I* opposite to *Q*. Hence *IP* and *IP'* are normals.

3. *The lemniscate.* If the curve is drawn by the method of Fig. 83 (p. 115), the instantaneous centre of *MM'* is at the intersection of *CM* and *C'M'*. The line drawn from this point to *P* is a normal to the curve.

*   4. *The right strophoid.* The following method is suggested: Let *OA* and *AD* be two lines at right angles, *O* being a fixed point about 2 in. from *A*. With centre at any point *W* on *OA* produced, and radius *WO*, draw an arc cutting *AD* at *Q*. With the same radius, and centres at *O* and *Q*, draw arcs intersecting at *T* (so that *OWQT*

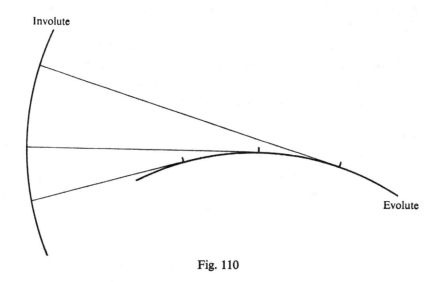

Involute

Evolute

Fig. 110

is a rhombus). Join *WQ* and mark off *WP* on it equal to *WA*. Join *PT*. Then the locus of *T* is a parabola; that of *P* is a strophoid; and *PT* is a normal to the strophoid. (*Note:* The position of *W* should be moved towards *A* and past *A*, until it is half-way between *A* and *O*; but, after it has passed *A*, *WP* must be marked off along *QW* produced.) (*Hint for proof:* In Fig. 66, p. 93, *T* is the instantaneous centre of the moving set square.)

Involutes

Every example of an evolute is also one of an involute: thus the catenary is the evolute of the tractrix and the tractrix is an involute of the catenary. The tangent to the evolute is the normal to the involute, and its length, measured between the two curves, is the radius of curvature of the involute (Fig. 110). As explained on p. 84, the difference in length between two of these tangents is equal to the length of arc of the evolute, measured between their points of contact. The involute may thus be thought of as the locus of a point of a string which is laid along the evolute and unwrapped.

To Draw an Involute

The involute of a given curve may be drawn approximately as follows: Draw a number of tangents to the given curve. With centre at the intersection of two

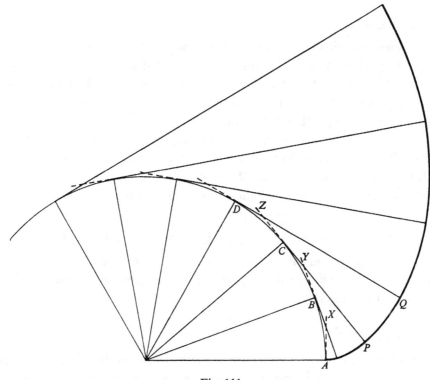

Fig. 111

neighbouring tangents draw an arc, bounded by those tangents, passing through the point of contact of one of them (Fig. 111). Repeat for the next pair of tangents, using such a radius as will make the arcs join; and so on.

The error in this method is due to the fact that the length of arc of the original curve is replaced by the sum of the segments of the tangents, which is necessarily more than the true length of the arc. But the error can be made as small as we please by taking the tangents near enough together.

Involutes of the Circle

* This curve, shown in Fig. 111, is commonly used for the shaping of cog-wheels. In Fig. 109, it is desired that two wheels, whose centres are at *A* and *B*, should revolve as if the two *pitch-circles*, in contact at the *pitch-point P*, were rolling against each other. *QPR* is drawn at a convenient angle, usually 20°, to the common tangent at *P*; *Q* and *R* being the feet of the perpendiculars from *A* and *B*. *Base-circles* are drawn with centres *A* and *B*, radii *AQ* and *BR*. The profiles of the teeth are then drawn as involutes of the two base-circles.

Suitable Dimensions

The radii of the pitch-circles may conveniently be in the ratio 4:3. The drawing is made somewhat easier if the angle between *QPR* and the common tangent is increased to 25°. To find the positions of successive teeth, mark off equal distances along the pitch-circles. The tops of the teeth should be arcs of circles concentric with, and slightly larger than the pitch-circles.

Proof: To see that, with the teeth in contact, the wheels will in fact revolve as if the pitch-circles were rolling against each other, consider two points Q' and R' which will move to the positions Q and R in the same interval of time. If the tangents to the base-circles at Q' and R' are $Q'Y$ and $R'Z$,

$$Q'Y + ZR' = QP + PR.$$

But $\qquad Q'Y = \text{arc } Q'Q + QP \quad$ and $\quad ZR' = PR - \text{arc } R'R \text{ (constr.)}.$

Therefore arc $Q'Q$ = arc $R'R$, and it follows that Q and R move with equal velocities. As the radii are in proportion, points fixed on the pitch-circles will also move with equal velocities.

Parallel Curves

While every curve has but one evolute, it has many involutes; for the initial point, where the involute cuts the original curve, may be chosen arbitrarily. The various curves so obtained are called *parallel curves*. Any two of them are a constant distance apart, the distance being measured along the common normal. The involutes of a circle are all identical, but in other cases varying shapes are produced. To draw curves parallel to a given curve it is only necessary to draw a number of normals and to mark off equal distances along each of them.

Drawing of Involutes and Parallel Curves

It may be noted that, in drawing involutes, a cusp may occur either at the initial point (i.e. the point where the involute meets the original curve) or at a point corresponding to a point of inflexion of the original curve. In drawing a curve parallel to a given curve, a cusp may be found at a point where the radius of curvature of the original curve is equal to the constant distances between the curves.

The following are suggested for drawing:

1. An involute of the circle.

2. Involutes of the nephroid (i) with the initial point mid-way between two cusps, (ii) with the initial point at a cusp-point.

3. An involute of the lemniscate, with the initial point at one end of the transverse axis.

4. Curves parallel to the parabola, with the constant distance measured along the inward normal and (i) equal to $2a$ (a being the distance from the focus to the vertex), (ii) greater than $2a$.

5. Curves parallel to the ellipse, with the constant distance measured along the inward normal and (i) between b^2/a and a^2/b, (ii) greater than a^2/b.

6. Curves parallel to the astroid, at varying distances.

Some Curves and their Evolutes

Curve	*Evolute*
Parabola	Semi-cubic parabola
Ellipse or hyperbola	Lamé curves, $(x/A)^{\frac{2}{3}} \pm (y/B)^{\frac{2}{3}} = 1$
Cycloid	An equal cycloid
Epicycloid or hypocycloid	A similar epicycloid or hypocycloid
Cayley's Sextic	Nephroid
Equiangular spiral	An equal spiral
Tractrix	Catenary

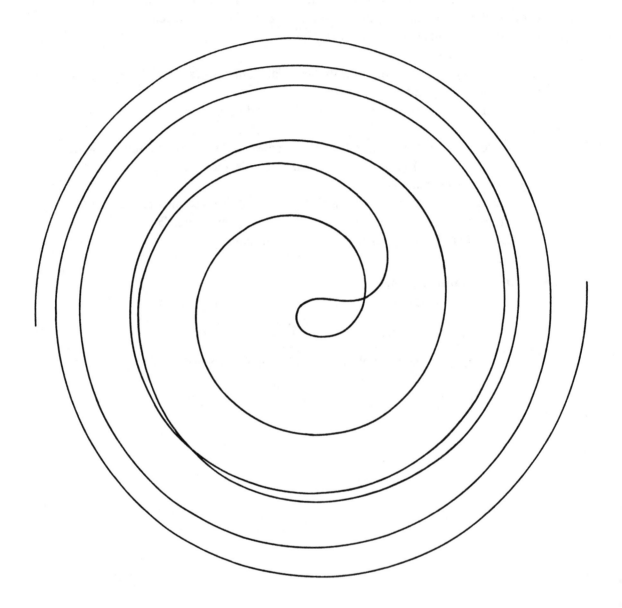

Fig. 112. A parabolic spiral

22

SPIRALS

Definition

The word *spiral*, in its mathematical sense, means, properly speaking, a plane curve traced by a point which winds about a fixed pole from which it continually recedes; but the use of the word has been extended to other curves, for example the so-called *sinusoidal spirals*, in which the tracing-point moves alternately towards and away from the pole.

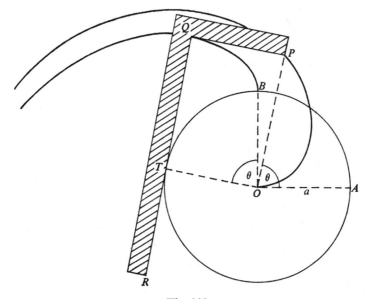

Fig. 113

The Spiral of Archimedes

This is a true spiral, defined by the polar equation $r = a\theta$. It can be easily drawn with the aid of polar graph paper. The successive whorls (i.e. circuits of the pole) are spaced at equal intervals.

It may also be drawn by rolling one arm of a carpenter's square along the circumference of a fixed circle whose radius is equal to the inside edge of the other

[173]

arm. Suppose that, in Fig. 113, P and Q have started from O and B respectively, and that $TQ =$ arc TB. Then Q describes an involute of the circle; and, since $OP = TQ =$ arc TB, OP is proportional to θ. Therefore P describes an Archimedean spiral. (If θ is measured in radians, $OP = a\theta$, and the polar equation of the spiral is $r = a\theta$.)

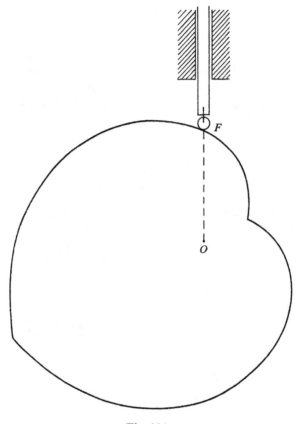

Fig. 114

As T is the instantaneous centre, the normal at P passes through T and can thus be drawn.

Used as a cam, this curve converts uniform angular motion into uniform linear motion. Fig. 114 shows the section of a heart-shaped cylinder in which each part of the curve is an Archimedean spiral with O as pole. The follower F can move in a vertical line through O. If the cam rotates about O with uniform angular velocity, F will move in reciprocating motion with uniform speed.

[174]

Other Spirals of the Family $r^n = a^n\theta$

These spirals are easily drawn on polar graph paper and several of them are of interest.

Fermat's spiral, $r^2 = a^2\theta$. There are *plus* and *minus* values of r for any positive value of θ; the curve, therefore, has central symmetry about the pole.

The reciprocal (or hyperbolic) spiral, $r\theta = a$. This curve has as asymptote a straight line distant a from the pole (θ being measured in radians).

The lituus, $r^2\theta = a^2$. The initial line is an asymptote. The curve may be defined as the locus of a point P such that the area of the circular sector whose bounding radii are OP and an equal length OQ along the initial line is constant.

If the curve is plotted for positive values of r only, the result is said to resemble the volute in the capital of an Ionic column.

Other Spirals

The equiangular spiral. This has already been discussed (ch. 11).

The parabolic spiral, $(r-a)^2 = b^2\theta$. If both values of r are plotted, an unending curve is produced, the two parts of it crossing and recrossing each other an infinite number of times (Fig. 112). (In drawing this curve any unit of angle may be chosen. Convenient dimensions are given by $a = 2$, $b = 1$, unit of angle $= 10°$.) Fermat's spiral is a special case.

The sinusoidal spirals, $r^n = a^n\cos n\theta$. These are not true spirals, in that the tracing-point does not continually recede from the pole. For different values of the parameter n we obtain a family of curves which includes the lemniscate ($n = 2$), the straight line ($n = -1$) and the rectangular hyperbola ($n = -2$). With fractional values we have the cardioid ($n = \frac{1}{2}$), the parabola ($n = -\frac{1}{2}$), Cayley's Sextic ($n = \frac{1}{3}$) and the Tschirnhausen Cubic ($n = -\frac{1}{3}$).

**   It may be proved by the methods of the calculus that the pedal equation of a sinusoidal spiral is $r^{n+1} = a^n p$; and hence that the pedal curve with respect to the pole is another sinusoidal spiral, with its parameter n' equal to $n/(n+1)$. Interchanging n and n' it follows that the negative pedal of a sinusoidal spiral with respect to its pole is another such curve, with parameter $n/(1-n)$.

Archimedes wrote a work *On Spirals* in which he proved that the polar subtangent of a point on his spiral was equal in length to an arc of a circle. In Fig. 113, if the tangent at P meets TO produced at L, the polar subtangent OL is equal to an arc of a circle drawn with centre O from P to the point where it meets OA produced. In this sense he rectified the circle.

Lituus means a 'crook', for example an augur's staff or a bishop's crosier; Maclaurin used the term in his *Harmonia Mensurarum* in 1722.

[175]

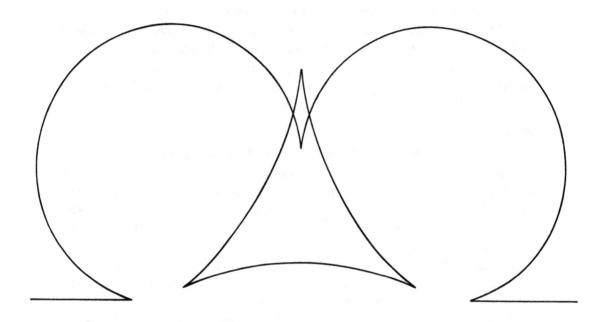

Fig. 115. The deltoid and its inverse with respect to the mid-point
of one of its arcs

23

INVERSION

Definition

Let S be a given curve and let O be a fixed point, with k a constant distance. If a radius vector OP is drawn from O to the curve, and if P' is a point on OP such that $OP.OP' = k^2$, then P' is said to be the *inverse of P with respect to O*; and, if the locus of P' is a curve S', then S' is said to be the *inverse of S with respect to O*.

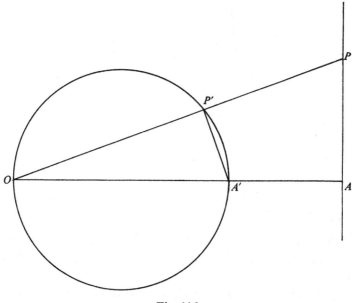

Fig. 116

It follows immediately that P is the inverse of P', and S of S'; the relationship is a mutual one. It may also be noted that the constant k determines the size of the inverse curve, not the shape. Variation of k would produce a set of similar curves. The circle whose centre is O and whose radius is k is called the *circle of inversion*; it is used in the geometrical theory of inversion but it is not of great importance for our purposes.

Examples of Inversion

Several examples have already been given; e.g. the cardioid has been drawn as the inverse of the parabola (p. 39) and the lemniscate as the inverse of the rectangular hyperbola (p. 115). It can also be seen from Fig. 116 that the inverse of a straight line is a circle through the centre of inversion: for, if OA is the perpendicular from O to the line, and A', P' are the inverses of A, P respectively, then A, A', P', P are concyclic and angle $OP'A' =$ angle OAP, a right angle. Therefore the locus of P' is a circle on OA' as diameter.

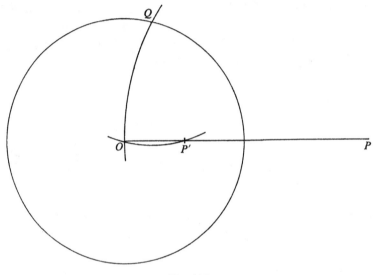

Fig. 117

Polar Equations

If (r, θ) are the polar coordinates of P, with O as pole, and (r', θ) are those of P' then $rr' = k^2$; and, if the polar equation of S is written as $r/k = f(\theta)$, that of S' is $k/r = f(\theta)$.

For example, among the spirals, the inverse of the spiral of Archimedes, $r = a\theta$, is the reciprocal spiral $r\theta = a$. In the same way the lituus and the spiral of Fermat are inverse curves; and the sinusoidal spirals arrange themselves in pairs of inverse curves, with n taking corresponding positive and negative values.

Drawing Inverse Curves

Method 1. Join points on the curve to the centre of inversion and use a table of reciprocals to plot the inverse points.

Method 2. Draw the circle of inversion, centre O (Fig. 117). With centre at P on

[178]

the given curve, and with radius *PO*, draw an arc cutting the circle at *Q*; with centre *Q* and radius *QO* draw an arc cutting *OP* at *P'*. Then *P* and *P'* are inverse points.

(The method fails for points near *O*, but it can, if necessary, be applied in reverse: i.e. given *P'*, draw the perpendicular bisector of *OP'* to meet the circle at *Q*, and that of *OQ* to meet *OP'* produced at *P*.)

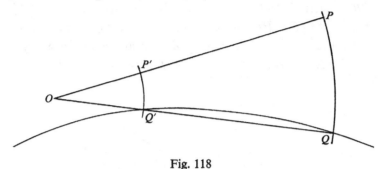

Fig. 118

Method 3. Draw a circle, of large radius, passing near *O* (Fig. 118). With centre *O* and radius *OP* draw an arc to cut the circle at *Q*. Join *OQ*, cutting the circle again at *Q'*. With centre *O* and radius *OQ'* draw an arc cutting *OP* at *P'*. Then *P* and *P'* are inverse points.

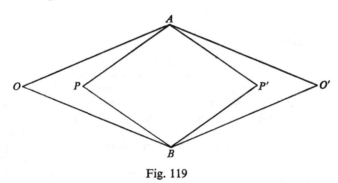

Fig. 119

Note: Whatever method is used, it is helpful to know that the original curve and the inverse make supplementary angles with any straight line through *O*.

Linkages for Inversion

Inverse curves may also be drawn by mechanical means:

1. *Peaucellier's cell.* In Fig. 119, the eight lines represent rods freely jointed together at their ends, *OAO'B* and *PAP'B* being rhombuses. If *O* is kept fixed,

P and P' are inverse points with respect to O. (*Hint for proof:* Consider the circle, centre A and radius AP, cutting OA and OA produced at C and D. Then C and D are at fixed distances from O and $OP.OP' = OC.OD$.)

Alternatively, if P is kept fixed, O and O' are inverse points.

2. *Hart's linkage.* In Fig. 120, $ABCD$ is a 'crossed parallelogram' of jointed rods (i.e. $AB = CD$ and $AD = BC$). O, P, P' are fixed points on the rods, dividing AB, AD, CB respectively in the same ratio. If O is kept fixed, P and P' are inverse points with respect to O. (*Hint for proof:* OPP' is a straight line parallel to AC and BD. Consider the circle $APP'C$ cutting AB, produced if necessary, at K. Then $BP'.BC = BA.BK$. $\therefore$ BK is constant. Now consider $OP.OP'$.)

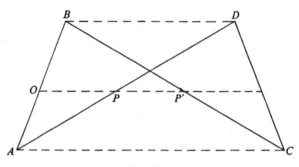

Fig. 120

Pairs of Inverse Curves

	Centre of inversion		
First curve	*as point of first curve*	*as point of second curve*	*Second curve*
Straight line	Point not on the line	Point on circumference	Circle through the point
Circle	Point not on circumference		Circle
Parabola	Focus	Cusp-point	Cardioid
Parabola	Vertex	Cusp-point	Cissoid of Diocles
Rectangular hyperbola	Centre	Centre	Lemniscate
Rectangular hyperbola	Vertex	Node	Right strophoid
Ellipse or hyperbola	Focus	Pole, or node	Limaçon $(r = k - 2a\cos\theta)$
Hyperbola with asymptotes at $60°$ to transverse axis	Vertex	Node	Trisectrix of Maclaurin
Spirals $(r^n = a^n\theta)$	Pole	Pole	Spirals $(r^n\theta = a^n)$
Sinusoidal spirals $(r^n = a^n\cos n\theta)$	Pole	Pole	Sinusoidal spirals $(r^n\cos n\theta = a^n)$

Some Further Suggestions for Drawings

1. Inverses of the cardioid (i) with respect to the point of the curve directly opposite the cusp, (ii) with respect to the centre of the base-circle.

2. Inverse of the parabola with respect to the point of intersection of the axis and the directrix.

3. Inverse of the lemniscate (i) with respect to one end of the transverse axis, (ii) with respect to a point on the conjugate axis.

4. Inverse of the cissoid of Diocles with respect to a point not on its axis of symmetry.

5. Inverse of the astroid with respect to one of its cusp-points.

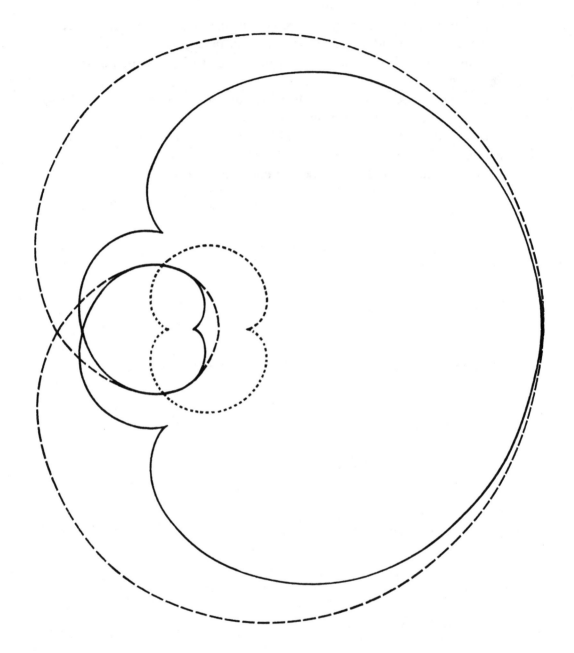

Fig. 121. The trisectrix, its evolute, and caustic with
radiant point opposite the node

24

CAUSTIC CURVES

Definition

Let S be a given curve and let F be a fixed point called the *radiant point*. If rays from F are reflected by the curve, the envelope of the reflected rays is called the *caustic of S with F as radiant point*. Expressing this geometrically, let any line through F meet the curve at Q and let QP be drawn so that QP and QF make equal angles with the tangent to the curve at Q; then the envelope of QP is the caustic.

This *caustic by reflection* is sometimes called the *catacaustic*, to distinguish it from a curve similarly formed (the *diacaustic*) when the rays are refracted. Diacaustics will not be discussed here and the word *caustic* will be used to mean the *catacaustic*.

The Drawing of Caustics

In general it is necessary to draw the normal (or the tangent) to the curve at Q. If the image of F in the normal (or the tangent) is joined to Q, this is the line QP whose envelope is the caustic. If the curve S is a circle, however, use can be made of the fact that equal chords drawn from a point on the circumference are equally inclined to the radius.

The following examples are suggested for drawing:

1. The caustic of a circle with radiant point on the circumference. This is the cardioid, as proved on p. 41 (see also below).

2. The caustic of a circle with radiant point at infinity. This is the nephroid, as shown on p. 70.

3. The caustic of a circle with radiant point inside or outside the circle.

4. The caustic of a parabola for parallel rays perpendicular to its axis (*Tschirnhausen's Cubic*).

5. The caustic of a cycloidal arch for parallel rays perpendicular to the base. (In Fig. 57, p. 83, it can be seen that the reflected ray is the radius PO of the rolling circle. See also p. 86, Ex. 2.)

6. The caustic of a cardioid with radiant point at the cusp. (In Fig. 24, p. 38, PQ is the normal. If the tangent to the circle at Q is drawn first, and P found as

the image of A in the tangent, the normal PQ can be accurately drawn. The resultant caustic is a nephroid.)

7. The caustic of a cardioid with radiant point opposite to the cusp.

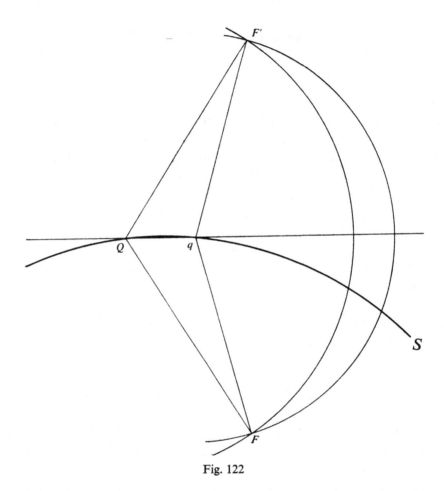

Fig. 122

The Caustic as an Evolute

* In Fig. 122, Q and q are points on the given curve and circles are drawn through F, intersecting at F', with these points as centres. By congruent triangles, F' is the image of F in Qq and, in the limit, as q approaches Q, the image of F in the tangent at Q to the given curve. F', moreover, becomes a point on the envelope of the circles, and $F'Q$ becomes a normal to this envelope. (It will be recalled that the cardioid, for example, was drawn as an envelope in this way.) Now the locus of F' is a curve similar to the pedal of the given curve with respect to F, but on double scale. (This curve is called the *orthotomic of S with respect to F*.) $F'Q$ is, therefore,

[184]

the normal to the orthotomic. But the envelope of $F'Q$ is the caustic. Hence the caustic is the evolute of the orthotomic.

The nature of certain caustics can be easily seen from this. For example, the pedal (or orthotomic) of a circle with respect to a point on its circumference is a cardioid; and the evolute of a cardioid is another cardioid; therefore the caustic of a circle with radiant point on the circumference is a cardioid. Again, the pedal of an ellipse with respect to its focus is a circle; and the evolute of a circle is a single point, the centre. Hence the caustic of an ellipse with radiant point at a focus is a single point (the other focus). By the same argument, the caustic of a rectangular hyperbola with radiant point at its centre is the evolute of the lemniscate.

Some Curves and their Caustics

Curve	Rays	Caustic
Circle	From point on circumference	Cardioid
Circle	Parallel	Nephroid
Parabola	Perpendicular to axis	Tschirnhausen's Cubic
Tschirnhausen's Cubic	From pole	Semi-cubic parabola
Cardioid	From cusp	Nephroid
Deltoid	Parallel (in any direction)	Astroid
Cissoid	From point on axis (beyond asymptote) whose distance from the cusp is four times that of the asymptote	Cardioid
Cycloidal arch	Perpendicular to base	Two cycloidal arches
Equiangular spiral	From pole	An equal spiral

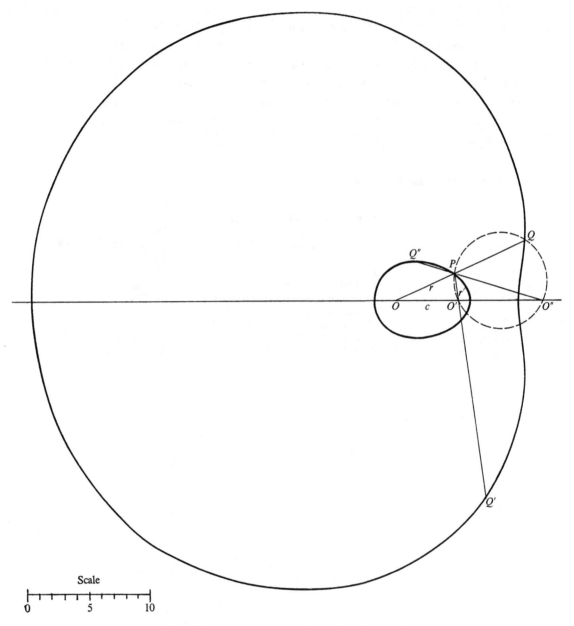

Scale

0   5   10

Fig. 123. Cartesian ovals. $3r \pm 2r' = 20$, $c = 5$

25

BIPOLAR COORDINATES

Definition

If the distances of a point P from two fixed points O and O' are r and r' respectively, then r and r' are called *bipolar coordinates* of P. An equation connecting r and r' may define a locus and is then called the *bipolar equation* of the locus. If the distance OO' is c, the three lengths r, r' and c (which are all supposed positive) must form a triangle; i.e. $r+r' > c$ and $-c < r-r' < c$.

As P may be on either side of OO', every locus defined by such an equation will be symmetrical about OO'.

Examples of Bipolar Equations

The ellipse can be defined by the equation $r+r' = 2a$; and the hyperbola by $r-r' = \pm 2a$. It has also been shown (p. 116) that the lemniscate has the bipolar equation $rr' = \frac{1}{2}a^2$, where $OO' = a\sqrt{2}$.

The Ovals of Cassini

These curves are defined by the bipolar equation $rr' = k^2$. Let $OO' = 2c$. Then, if $k = c$, we have the lemniscate as a special case. If $k > c$, the curve is a single oval; if $k < c$, it separates into two. A convenient way of drawing the curve is to draw a circle of radius c and mark a point A outside it such that the tangent from A to the circle is of length k. If a line through A meets the circle at Q and Q', then AQ and AQ' are of length r and r'. (The extreme width of the oval, measured along its axis, will be twice the distance from A to the centre of the circle.)

The normal may be drawn from the fact that, if C is the centre and P a point on the curve, the normal and PC make equal angles with PO and PO'.

** *Hint for proof:* By applying the sine formula to triangles CPO and CPO' it can be proved that
$$\frac{\sin CPO}{\sin CPO'} = \frac{r'}{r}.$$

If SPT is the tangent at P, and PG the normal,
$$\frac{\sin GPO}{\sin GPO'} = \frac{\cos OPS}{\cos O'PT}.$$

[187]

But $\cos OPS = dr/ds$ and $\cos O'PT = -dr'/ds$, and the ratio of these quantities may be found by differentiating the equation $rr' = k^2$.

The Ovals of Descartes

* These curves are defined by the linear relations

$$mr \pm nr' = k.$$

They occur in conjugate pairs, as indicated by the $\pm$ sign. To draw them, it is convenient to plot first the graphs of the straight lines $mx \pm ny = k$ (Fig. 124a). The coordinates (x, y) of any point on one of these lines give values of r and r' satisfying the above equation, but the requirement that r, r' and c (where $OO' = c$) must be sides of a triangle restricts the choice to points within the rectangle marked out by dotted lines. Let P be any such point. With O and O' (Fig. 124b) as centres, and with radii equal to the x- and y-coordinates of P respectively, arcs are drawn to intersect at a point on the required curve.

It may be noted that points on the perimeter of the rectangle correspond to points on the axis of symmetry of the curve; points on the short side to points between O and O', those on the other two sides to points on the left of O and on the right of O' respectively.

Focal Properties

** The polar equation, with O as pole and OO' as initial line, is obtained by eliminating r' from the above equation and

$$r'^2 = r^2 + c^2 - 2rc\cos\theta,$$

where angle $O'OP = \theta$. The result, whether the *plus* or *minus* sign is taken, is

$$r^2(m^2 - n^2) + 2r(cn^2\cos\theta - km) + (k^2 - n^2c^2) = 0.$$

The roots of this equation are the lengths of OP and OQ; these will be called r and R respectively. To obtain an equation for the two points in which QO produced meets the ovals again it would be necessary to change θ into $180° - \theta$. The product of the roots, rR, is equal to $(k^2 - n^2c^2)/(m^2 - n^2)$, a constant, showing that the two curves are inverses with respect to O.

If angle $OO'P$ is called ϕ, a similar elimination of r gives

$$r'^2(n^2 - m^2) + 2r'(cm^2\cos\phi \mp kn) + (k^2 - m^2c^2) = 0.$$

The roots of this equation, which will be called r' and R', are the lengths of $O'P$ and $O'Q'$, where Q' is either the point where $O'P$ meets the same oval again (e.g. the inner oval in Fig. 124b) or the point where PO' produced meets the other oval of the pair (e.g. the outer oval in Fig. 123). (This apparent anomaly is due to the fact that $\cos\phi$ changes sign for a radius drawn in the opposite direction.) The two cases are

[188]

distinguished by the sign of the product $r'R'$. Fig. 124 is drawn with $k/m < c < k/n$, and both the products rR and $r'R'$ are positive. In Fig. 123, k/m and k/n are both greater than c, with m greater than n; the product rR is again positive, but $r'R'$ is negative.

Now let a circle be drawn through Q, P and O', meeting OO' again at O'' (the inner oval in Fig. 123, the opposite side of the outer oval in Fig. 124b).

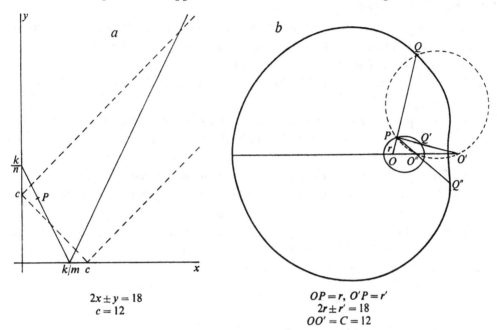

$$2x \pm y = 18$$
$$c = 12$$

$$OP = r,\ O'P = r'$$
$$2r \pm r' = 18$$
$$OO' = C = 12$$

Scale: 1·5 mm. to 1 unit.

Fig. 124

If $O'Q$ is joined, triangles OPO'' and $OO'Q$ are similar, and hence

$$\frac{O''P}{OO''} = \frac{O'Q}{OQ}.$$

But $O'Q$ is related to OQ by the equation

$$m.OQ \mp n.O'Q = k$$

(the upper sign being taken when Q is on the outer branch, as shown in both diagrams).

$$\therefore \quad \frac{O''P}{OO''} = \frac{n.O'Q}{n.OQ} = \pm\frac{m.OQ - k}{n.OQ} = \pm\left(\frac{m}{n} - \frac{k}{n}\cdot\frac{m^2 - n^2}{k^2 - n^2 c^2}\,OP\right).$$

Hence r'' and r are connected by a linear relation. Thus, if O and O' are called foci of the curve, O'' is a third focus, in the sense that it could be used with O for plotting the curve from a bipolar linear equation. The ovals shown in Fig. 123

[189]

are in fact the same as those in Fig. 124 (but on a larger scale), the foci O' and O'' being interchanged.

It follows, too, that there is a linear relation connecting r' and r'', so the three foci, O, O' and O'' are on a par with each other. If the pair of ovals is regarded as one curve, that curve can be inverted into itself with respect to any one of these foci.

Some Further Examples

1. The equation $r = kr'$ represents a circle; and, if k is varied, a set of coaxial circles with limiting points at O and O'. (*Hints for proof:* Let the locus cut the axis at A and A'. Then $OA/AO' = OP/PO' = OA'/O'A' = k$. Therefore PA and PA' are the bisectors of angle OPO'. For the second part, let $CA = x$ and $CA' = x'$; prove that $xx' = \frac{1}{4}c^2$.)

2. The equation $lr^2 + mr'^2 = c^2$ represents a circle whose centre Q divides OO' in the ratio $m:l$. (*Hint for proof:* Apply the cosine rule to triangles OQP and $O'QP$.) Consider the special cases $l = m = 1$, $l = m = 2$ and $l = -m$.

3. The equation $r^2 = ncr'$, where $OO' = c$ and n is positive, represents a set of ovals. Plot these curves for $n = \frac{1}{2}$, 1, 2, 3, 4, 5 (c may conveniently be taken as 1 unit). Draw tangents from O'. Use the cosine rule to investigate further.

** 4. The reflecting property of the ellipse may be proved from the bipolar equation $r + r' = 2a$. If the angles made by the curve with OP and $O'P$ are ϕ and ϕ' respectively (measured in opposite senses),

$$\cos\phi = \frac{dr}{ds} \quad \text{and} \quad \cos\phi' = -\frac{dr'}{ds}.$$

But $dr/ds + dr'/ds = 0$; therefore $\phi = \phi'$.

** 5. Prove the reflecting property of the hyperbola.

** 6. Prove that, for the ovals of Cassini, if lines drawn from O and O' at right angles respectively, to OP and $O'P$ meet the tangent at P in the points T and T', then $TP = PT'$.

** 7. Let P be a point of the locus $r = kr'$. Use the method of no. 4 to prove that lines drawn from O and O' at right angles to OP and $O'P$ respectively meet the tangent at P at the same point. Hence find another way of proving the second part of no. 1.

** 8. *Magnetic lines.* Suppose equal and opposite magnetic poles are at O and O'. For lines of magnetic force, resolving along the normal, $\mu \sin \phi / r^2 - \mu \sin \phi' / r'^2 = 0$. If angle $O'OP = \theta$ and angle $OO'P = \theta'$, so that $r \sin \theta = r' \sin \theta'$, prove that $\cos \theta + \cos \theta' = $ constant.

For equipotential lines, since no work is done as a pole is moved along the line, $\mu \cos \phi / r^2 - \mu \cos \phi' / r'^2 = 0$; therefore $1/r - 1/r' = $ constant.

[190]

FURTHER READING

Useful catalogues of curves and their properties, without proofs, are to be found in:

R. C. Archibald, *Encyclopaedia Britannica, SPECIAL CURVES*;

R. C. Yates, *Curves and their Properties* (Michigan, 1947).

The drawing of some of the curves, by geometrical and mechanical means, is dealt with in:

H. M. Cundy and A. P. Rollett, *Mathematical Models* (Oxford, 1951);

R. C. Yates, *A Mathematical Sketch and Model Book* (Louisiana, 1941);

F. C. Boon, *A Companion to Elementary School Mathematics* (London, 1924);

A. B. Kempe, *How to Draw a Straight Line; a Lecture on Linkages* (London, 1877, *reprinted* New York, 1953).

Certain groups of curves are fully discussed in:

R. A. Proctor, *The Geometry of Cycloids* (London, 1878, *out of print*);

W. H. Besant, *Roulettes and Glissettes* (Cambridge, 1870, *out of print*).

Some treatment of curves will also be found in most calculus books, notably in:

J. Edwards, *Differential Calculus* (London, 1893);

H. Lamb, *Infinitesimal Calculus* (Cambridge, 1897);

C. V. Durell and A. Robson, *Elementary Calculus*, vol. II (London, 1934);

A. W. Siddons, K. S. Snell and J. B. Morgan, *A New Calculus*, Part III (Cambridge, 1952);

E. A. Maxwell, *An Analytical Calculus*, Vols. II, III (Cambridge, 1954).

Special curves are treated more comprehensively, with historical notes, in:

G. Loria, *Spezielle Algebraische und Transzendente Kurven* (Leipzig and Berlin, 1902, *out of print*);

F. G. Teixeira, Vols. IV, V. *Courbes Speciales Remarquables* (Coimbre, 1907, *out of print*);

H. Wieleitner, *Spezielle Ebene Kurven* (Leipzig, 1908, *out of print*).

The tracing of curves from their equations is more fully dealt with in:

P. Frost, *Curve Tracing* (London, 1892).

The general treatment of plane curves by advanced methods will be found in:

G. Salmon, *Higher Plane Curves* (Dublin, 1879, *reprinted* New York, 1960);

H. Hilton, *Plane Algebraic Curves* (Oxford, 1932, *out of print*);

E. J. F. Primrose, *Plane Algebraic Curves* (London, 1955).

GLOSSARY

Abscissa. See *Cartesian coordinates.*

algebraic curve. One whose Cartesian equation is algebraic, i.e. can be expressed by rational powers of x and y connected by the operations of addition, subtraction, multiplication and division; e.g. $y^2 = x/(x+y)$, but not $y = 2^x$.

analytical (or coordinate) geometry. The geometry of Descartes, in which a point is a pair of numbers (x, y), and a curve is an equation connecting x and y. See *Cartesian coordinates.*

asymptote. The limiting position of a tangent to a curve as the point of contact moves indefinitely far from the origin; see p. 25.

axes of coordinates. Fixed lines with reference to which the positions of points, lines and curves are specified. See *Cartesian coordinates.*

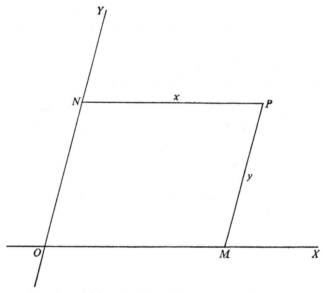

Fig. 125

axis of a curve. An axis of symmetry for the curve; i.e. a straight line such that, if P is any point of the curve, the image of P in the line is another point of the curve.

bipolar coordinates. See p. 187.

Cartesian coordinates. Those invented by Descartes. The position of a point P in a plane is related to two fixed axes, OX and OY (Fig. 125), the position being specified as (x, y), where x is the distance NP measured parallel to OX, and y is the distance MP measured parallel to OY. x is called the *abscissa*, or x-coordinate, of the point, and y is called the *ordinate*, or y-coordinate. If the axes are at right angles, the coordinates are said to be *rectangular*. The equation connecting x and y for points on a curve is called the *Cartesian equation* of the curve.

[192]

caustic. See p. 183.

cissoid. See p. 131.

conchoid. See p. 127.

conic sections. See p. 8.

curvature. Briefly, the circle of curvature at a point P of a curve is the circle that fits most closely to the curve at that point. The centre of curvature and the radius of curvature are the centre and radius of that circle. More accurately, if p is another point of the curve, and the normals at P and p intersect at C, the centre of curvature at P is the limiting position of C as p approaches P; and the radius of curvature is the distance from P to that limiting position.

cusp. If we suppose that a curve is traced by a moving point, a cusp-point is one where the moving point reverses its direction; and the form of the curve in such a neighbourhood is a 'cusp'; see Fig. 126, and Fig. 91, p. 138.

envelope. A curve touching every member of a system of lines or curves. See p. 3.

evolute. The locus of the centres of curvature at points of a curve; alternatively, the envelope of the normal to a curve. See Fig. 1, p. 2, and ch. 21, p. 167.

function. When a variable quantity y depends on another variable x in such a way that for certain values of x there are corresponding values of y, then y is said to be a 'function' of x. In Cartesian geometry this dependence may be illustrated by the drawing of a graph, the resultant curve then representing the functional dependence; e.g. the relationship $y = x^2$ is represented by the parabola of which it is the Cartesian equation.

image. If the perpendicular PM from a point P to a given line l is produced to P' so that $PM = MP'$, P' is called the 'image of P in the line l'.

inflexion. A point of inflexion on a curve is one where the tangent crosses the curve; e.g. in Fig. 56, p. 82, the dotted curve has points of inflexion where it crosses the upper 'horizontal' line.

initial line. See *polar coordinates*.

instantaneous centres. See pp. 48, 55.

intrinsic equation. If s is the arc-length measured from a fixed point on a curve to a variable point P, and ψ is the inclination of the tangent at P, measured from some fixed direction, the equation connecting s and ψ is called the 'intrinsic equation' of the curve; e.g. p. 121.

Fig. 126

inversion. See p. 177.

involute. See p. 167.

limit. If a variable point q moves towards and closely approaches a fixed point Q; and at the same time a point p, dependent on q, moves towards and closely approaches a fixed point P; then P is said to be 'the limiting position of p as q approaches Q'. E.g. in Fig. 3, p. 4, P is the limiting position of both p and P' as q approaches Q.

Similarly we may speak of the 'limiting position of a line', the 'limiting magnitude of an angle' and so on.

locus. Briefly, the path of a moving point; more accurately, the set of all possible positions which a point satisfying some condition can occupy.

negative pedal. If a curve S' is the pedal of another curve S with respect to a point O, then S is said to be the negative pedal of S' with respect to O. See also p. 157.

node. A point where two branches of a curve cross.

normal. The normal at a point P of a curve is the line through P at right angles to the tangent at P.

ordinate. See *Cartesian coordinates*.

origin. The point of intersection of a pair of Cartesian axes.

orthogonal. Crossing at right angles; i.e. for curves, having tangents at right angles at the point of intersection. An *orthogonal trajectory* is a curve crossing each of a given system of curves at right angles. The *orthogonal projection* of a curve on another plane is the locus of the feet of the perpendiculars from points on the curve to that plane.

orthotomic. The orthotomic of a curve with respect to a point F is the locus of the image of F in tangents to the curve; see pp. 184, 153.

parameter. A number used to specify a point on a curve (or, sometimes, one curve of a system). Often the Cartesian coordinates of the point are expressed in terms of a letter, say t, representing the parameter. These expressions are called the *parametric equations of the curve.* E.g. p. 82.

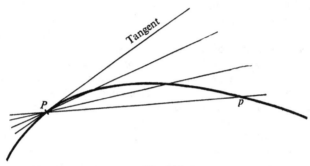

Fig. 127

pedal curve. The pedal curve (or simply 'the pedal') of a given curve with respect to a point F is the locus of the feet of the perpendiculars from F to tangents drawn to the given curve. See p. 153.

pedal equation. If P is a point on a curve and O is a fixed point (called the 'pole'), the length OP is called r and the perpendicular distance from O to the tangent to the curve at P is called p. The equation connecting p and r is variously described as the 'pedal equation' of the curve, the 'tangential-polar equation', or the 'p, r equation'. See p. 15.

polar coordinates. The position of a point P in a plane may be specified by (r, θ), where r is the distance of P from a fixed point O (called the 'pole') and θ is the angle made by OP with a fixed line (called the 'initial line'). The equation connecting r and θ for points on a curve is called the *polar equation* of the curve.

pole. A fixed point from which distances can conveniently be measured; usually in connection with polar or pedal equations, but also for such curves as spirals (p. 173) and strophoids (p. 135).

quadrature. Finding the area enclosed, or partly enclosed, by a curve. (Originally, drawing a square equal to that area.)

radius vector. A line drawn from a pole O to a variable point P on a curve.

rectangular axes. See *Cartesian coordinates.*

rectangular hyperbola. A hyperbola whose asymptotes are at right angles. See p. 26.

rectification. Finding the length of an arc of a curve. (Originally, drawing a right, i.e. straight, line equal in length to the arc.)

right circular cone. A cone on a circular base having its axis at right angles to the base.

similar. Figures are geometrically similar if to every point of one there corresponds a point of the other, corresponding angles being equal and corresponding lengths in a constant ratio.

GLOSSARY

spiral. See p. 173.

strophoid. See p. 135.

subtends. Is opposite to. More precisely, if P and Q are points of a line or curve, and A is another point, not on the line or curve, the angle PAQ may be described as 'the angle subtended at A by PQ'.

symmetry. A curve is said to be symmetrical about a line (the *axis of symmetry*) if the image in that line of every point on the curve is another point of the curve.

    A curve is said to be symmetrical about a point (the *centre*) if the image in that centre of every point on the curve is another point of the curve.

tangent. The tangent at P to a curve is the limiting position of a chord Pp, as p approaches P. P is called the *point of contact* of the tangent (Fig. 99).

touch. A line and a curve are said to touch each other if the line is a tangent to the curve. Two curves are said to touch each other if they have a common point and a common tangent at that point.

transcendental. A transcendental curve or equation is one which is not algebraic; e.g. $y = 2^x$, $y = \sin x$.

variable. A variable point is, briefly, a movable point; more accurately, it is a set of points satisfying some condition. A 'variable line' is similarly defined. In algebra, a letter such as x, which may have various numerical values, is called a 'variable'.

vertex. A point, other than a node, where a curve crosses an axis of symmetry. See pp. 4, 96.

INDEX OF NAMES

INDEX OF NAMES

INDEX OF SUBJECTS

The letter G suggests reference to the Glossary, S to the summaries at the ends of chapters. Numerals refer to pages.

INDEX OF SUBJECTS